Science and Complexity

British Library Cataloguing in Publication Data

Science and complexity.
1. Physics
I. Nash, Sara
530 QC21.2

ISBN 0-905927-32-X

Published by Science Reviews Ltd.,
40 The Fairway, Northwood, Middlesex, England.

Typeset by Geoffrey Morton MIOP
Brightwell, Wallingford, Oxfordshire.

Printed by Henry Ling Ltd., The Dorset Press,
Dorchester, England.

Science and Complexity

Edited by Sara Nash

Proceedings of an Interdisciplinary IBM Conference
London, February 1985

IBM United Kingdom Ltd.

Science Reviews Ltd.
1985

Foreword

The current volume represents the Proceedings of the fifth interdisciplinary conference held in London, England, under the auspices of IBM United Kingdom Ltd. with the general title Science and . . . . . Previous conferences took as their theme Science and Imagination, Science and the Unexpected, Science and Synergy, Science and Uncertainty.

Disparate enough topics, argues the conventional scientist, feeling even more uncomfortable when faced with the varied subjects discussed under each thematic umbrella, which encompass multi- and interdisciplinary domains of the sciences and even reach out towards the arts and the less-than-establishment thinking in the conventional sciences.

In the country of Victorian single science departments this will never do. But that is not what these conferences are about. In the same way as the progenitors of computers themselves tended to be unconventional scientists, the aim of the conferences is to bring together a wide range of specialists and offer them an opportunity to talk and argue with each other and indeed with the speakers. And it also harks back to another quality of science which seems to have been lost in an era of publish or perish: that science can be and often is fun, it is a challenging and entertaining set of mental gymnastics: and it is often when we do not consciously demand immediate usefulness from a piece of scientific research that the most valuable long term results will be obtained.

Such arguments are not to everybody's liking and it is hardly surprising that the conferences have received both high praise and strong criticisms. It should be said that IBM do not receive any direct benefit from underwriting this series of conferences: more praise then that they should perceive the long term benefits represented by the development of new and exciting initiatives from the rubbing together of differing and often antithetical exponents of the sciences and their followers.

The Publishers

Contents

Leslie Banks

Leslie Banks is Support Manager in IBM UK Academic Systems Marketing, and has chaired the annual interdisciplinary conference series, from 1981.

He joined IBM in 1963, after service in Local Government, Royal Artillery, Royal Air Force, Royal Navy, and United States Air Force.

Occasional spells with the University of London, the first attempt at an academic life in chemistry being interrupted by World War II, followed much later by BSc mathematics qualification and post-graduate diploma in computing science and numerical analysis.

His various IBM posts have included work involving research and academic institutions.

DFC, AFC, Croix de Guerre
Fellow, Royal Institute of Navigation (invention of navigation techniques)
Member of RGS Expeditions Committee, under Lord Schackleton, 1961–62.
Member, Royal Institution, and Society for Psychical Research.
Publications: "Pressure Pattern Flying" RIN; "Polar Air Navigation Guide" RAF; "Grid in all Latitudes" USAF; "An Aspect of Information Theory" IBM.

Introduction

Complexity may derive from an abundance of simplistic elements—the emergence of new properties which are not attributes of individual elements. The macro properties of physical objects, and our perceptions of those objects, are of quite different nature from the properties of microphysical entities. This volume is based on contributions to the IBM interdisciplinary conference *Science and Complexity—an Exploration of Complexity, with Simplicity as a Special Case.*

Simplicity has been the guiding principle of reductionism. In the wider sense, also, our search for order assumes that simplicity lies somewhere at the heart of natural complexity. This assumption has served us well until this century, but increasingly our searches reveal underlying, unexpected, complexities in Nature. The challenge is to map the observed complexities as a behavioural aspect of 'simple' elements.

In outward appearance, beautifully simple structures unfold in complex ways, and we find striking similarities between the self-organizing behaviour of seemingly quite different systems throughout the entire range from microphysics to cosmology, embracing man's world within that scale—in physics, chemistry, biology, sociology, ecology and planetary studies.

The symbol of the conference was the equiangular spiral. The spiral is manifest in the behaviour of certain insects: because of the way their eyes are constructed, they cannot look straight ahead. To attain an object—a candle for example—a moth will pursue a spiral path, an equiangular spiral, only to die in the moment of triumph. There is a parallel here with some scientific research, the painful, gradual gaining of a new 'truth'. But that is better than total paradigm blindness, or the headlong flight away from facts, from reality. As Paul Davies puts it: "however astonishing and inexplicable a particular occurrence may be, we can never be absolutely sure that at some distant time in the future a natural phenomenon will not be discovered to explain it." Or, as Niels Bohr phrases it, "is your suggestion sufficiently crazy?"

As practitioners in all disciplines have discovered, new lines of research

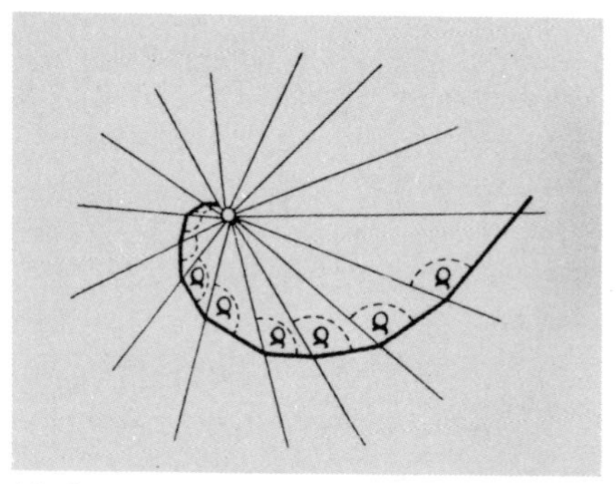

Moth converging on candle.

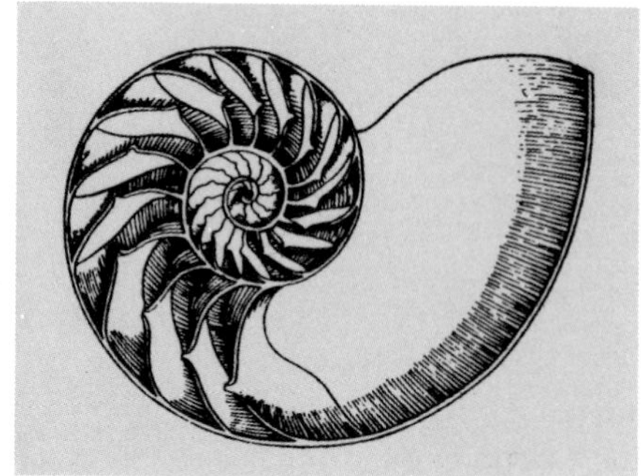

Nautilus.

follow the random walk described by François Jacob in *The Possible and the Actual*. "There is no way of telling", he wrote, "where a particular line of research will lead . . . this is why it is not possible to select some parts of science and to reject others. Either you have science or you don't have it. And if you have it you cannot take only what you like. You have to accept as well the unexpected and disturbing results". This conference explored some of those disturbing results.

Absolute certainties are no longer within the grasp of physicists, nor of biologists. Many biologists believe that the number of parameters within natural systems is not enough to explain the rich diversity of form, structure and function, including the cognitive functions of intelligent life. We have to take into account interactions with the environment along a time axis, in other words the unfolding of a complex universe reflected in the living world.

The unfolding equiangular spiral is manifested in the development of a wide range of natural systems. Emergent from its simple mathematical seed, the curve has the same form at all scales of growth. For example, the mollusc *Nautilus* grows its house in stages, based on the equiangular spiral. The shell never changes its shape—no middle-age spread—for this spiral is invariant under magnification. As Jacob Bernouilli's tombstone, featuring the formula for the spiral reminds us, "*Eadem mutata resurgo*", although changed I shall arise the same.

An equiangular spiral framed within its natural rectangle will contain approximate arcs of embedded squares which necessarily produce that spiral. That rectangle is the *Golden Rectangle*, with length and side in the *Golden Ratio*. The seed at the centre, that is the point of convergence of the squares, is a microcosm of the developed spiral. If *Nautilus* grew square chambers, at every stage of completion of a new chamber it would look like that rectangle.

The Golden Ratio is illustrated by the work of the sixteenth century Italian mathematician Fibonacci, also known as Leonardo of Pisa, famous

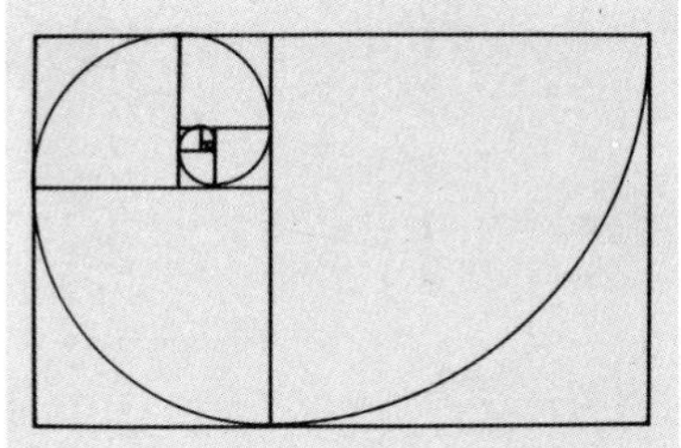

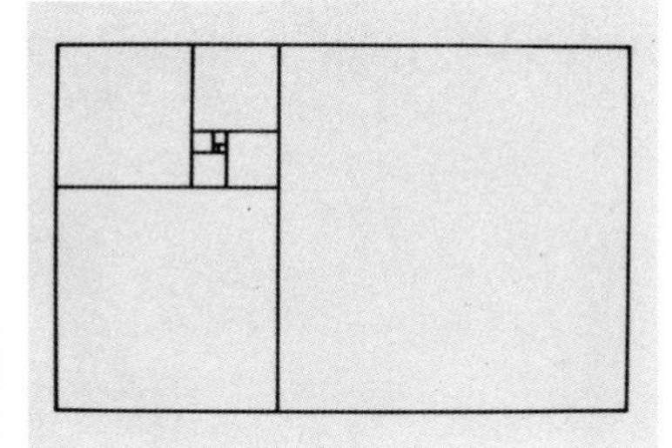

An equiangular spiral framed within its natural rectangle, with length and side in the *Golden Ratio*.

for the series consisting of a sequence where each number is equal to the sum of the previous two numbers . . . 1,1,2,3,5,8,13 . . .

from which we derive the ratios 1/1,2/1,3/2,5/3,8/5,13/8 . . .

which in the limit converge to the Golden Ratio 1.618 . . .

Long before Fibonacci, Nature discovered the benefits of the equiangular spiral. For example, it would not do for a sunseeking leaf to cover up another leaf lower down the stem: Nature invented a spiral distribution—phyllotaxis. Typical arrangements include the fifth leaf appearing after the third turn, or the eighth leaf after the fifth turn, and so on, to produce near optimum distributions for energy sharing. At least one species unfolds its leaves with separations of about 137 degrees, which subtends in Golden Ratio the residue of the 360 degree rotation, the spiral arrangement then providing the most efficient, democratic, sharing of energy from above, illustrating the inherent development of complex organization in the interaction of a system with its environment.

This conference explored the emergence of complexity in Nature, and the use of intellectual and technological tools in our search for order and explanation.

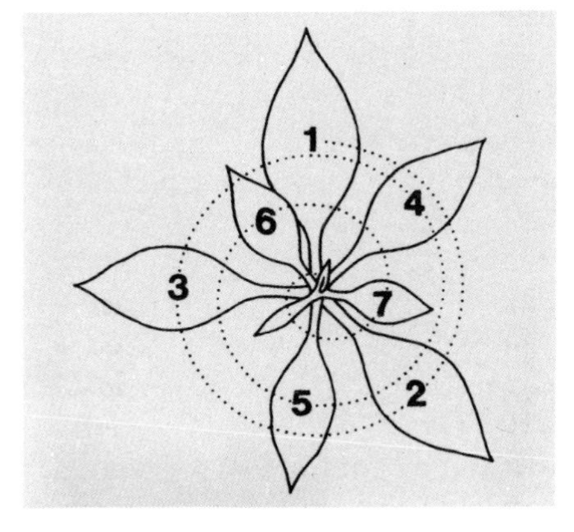

Optimal phyllotaxis, viewed from the top.

Professor Ilya Prigogine

Ilya Prigogine is Director, Instituts Internationaux de Physique et de Chimie, at the Free University of Brussells. He is also Director of the Statistical Mechanics and Thermodynamics Center of the University of Texas, Austin.

He achieved Docteur en Sciences Chimiques in 1941 from the Free University and continued his work there, leading to his present appointment in 1959. He has also served as Professor, Enrico Fermi Institute of Nuclear Studies, University of Chicago, and achieved recognition as Doctor Honoris Causa from twelve Universities in Belgium, France, Britain, USA, Sweden, Brazil and Poland. His membership of learned societies, and awards for achievement, moreover, extend also to Germany, India, Japan, USSR, Switzerland, Austria, and Rumania.

In 1955, the publication of "Thermodynamics of Irreversible Processes" was a significant treatment of systems far from equilibrium, which he called "dissipative structures". In 1977 he was awarded the Nobel Chemistry Prize for his work on "non-equilibrium thermodynamics", particularly his theory of dissipative structures.

The rediscovery of time

In the classical perspective, there was a clearcut distinction between what was considered to be simple and what had to be considered as complex: there was no hesitation about calling as 'simple' the Newtonian laws of motion, perfect gas, or chemical reactions. Also, one would have called as 'complex', biological processes, and even more so, human activities such as described by economics or urban planning. In this perspective, the aim of classical science was to discover, even in complex systems, some underlying simple level. This level would be the carrier of deterministic (such a carrier would be wave functions in the case of quantum mechanics) and time-reversible laws of nature: *future* and *past* would have the same role. However, this basic simple level remained elusive.

Today, a far reaching reconceptualisation of science is going on. Wherever we look, we find evolution, diversification and instabilities. We have long known that we are living in a pluralistic world in which we find deterministic as well as stochastic phenomena, reversible as well as irreversible. We observe deterministic phenomena such as the frictionless pendulum or the trajectory of the Moon around the Earth; moreover, we know that the frictionless pendulum is also reversible.

But other processes are irreversible, such as diffusion or chemical reactions; and we are obliged to acknowledge the existence of stochastic processes if we want to avoid the paradox of referring the variety of natural phenomena to a program printed at the moment of the big bang. What has changed since the beginning of this century is our estimation of the relative importance of irreversibility versus reversibility, of stochasticity versus determinism.[1–3]

Let us consider an example: the long-term variation of climate. We know that climate has fluctuated violently in the past. Climatic conditions that prevailed during the past 200 or 300 millions years were extremely different from what they are at present. During this period, with the exception of the Quaternary era (which began about 2 millions years ago) there was practically no ice on the continents, and the sea level was higher than its present value by about 80 metres. A striking feature of the Quaternary era is the appearance of a series of glaciations, with an average periodicity of 100000 years, on which is superposed an important amount of 'noise'. What is the source of these violent fluctuations which have obviously played an important role in our history? There is no indication that the intensity of the solar energy may have been responsible.

The temporal variation of climate is typically a 'complex process'. Again, in the perspective of classical physics, we would be tempted to attribute this complexity to a basic level, involving a large number of variables, which would enter into the determination of temperature. The situation would then be similar to that induced by the law of 'large numbers', which leads to fluctuations distributed in a Gaussian manner.

Recent progress in the study of the behaviour of dynamical systems enables us to determine the number of independent variables linked through differential equations whose solution could generate the observed temporal sequence of temperature. The unexpected outcome of this analysis[4] is that the number of independent variables which determine the climate, is only four. Therefore, we can no more ascribe the complexity observed to some underlying level, which would involve a large number of hidden variables. On the contrary, we have to attribute to the climatic system an *intrinsic* complexity and *impredictability*.

In a quite different field, recent work[5,6] has shown that the electrical activity of the brain in deep sleep as monitored by electroencephlogram (EEG) may be modelled by a fractal attractor. Deep-sleep EEG may be described by a dynamics involving five variables; again, this is very remarkable as it shows that the brain acts as a system possessing intrinsic complexity and unpredictability.

It is this instability which permits the amplifications of inputs related to sensory impression in the awake state. Obviously, the dynamical complexity of the human brain cannot be an accident. It must have been selected for its very instability. Is biological evolution the history of dynamical instability, which would be the basic ingredient of creativity characteristic of human existence?

There have been other surprises. Even in some of the simplest examples of dynamics such as an elastic pendulum, unexpected complexity has been discovered,[7,8] as it had been in some simple chemical reactions. It now seems that the gap between 'simple' and 'complex', between 'disorder' and 'order' is narrower than was thought before.

Complexity is no longer limited to biology or human sciences: it is invading the physical sciences and appears as deeply rooted in the laws of nature.

These new developments are likely to be of decisive importance. For many scientists, the unknown was lying only at the frontiers of physics: in cosmology and in elementary particles. Today, the interest in macroscopic physics and chemistry, dealing with phenomena on our own scale, is rapidly increasing. Let me present three reasons which, I believe, explain this interest.

(1) It leads to number of potentially innovative technical applications as well as to a better understanding of the main characteristics of our biosphere.

(2) It gives us the possibility of transferring the new theoretical tools coming from mathematical physics to biology and human sciences. It makes, therefore, the traditional distinction between hard and soft sciences obsolete.

(3) The basic characteristics of complexity are irreversibility and stochasticity. These concepts now begin to diffuse into the fundamental level of the description of nature.

Technical applications

The first science dealing with complexity in the field of physics and chemistry was the science of thermodynamics. The basic law of thermodynamics is the so-called 'Second Law', which expresses that entropy increases in isolated systems. (For more details, see refs 1–3 and 8.)

For a long time, the interest of thermodynamics concentrated on isolated systems at equilibrium. Today, interest has shifted to non-equilibrium, to systems interacting with their surroundings through an entropy flow. This interaction means that we are dealing with 'embedded' systems. This immediately brings us closer to objects such as towns or living systems, which can only survive because of their embedding in their environment.

There is another basic difference with classical mechanics. Suppose we have some foreign celestial body approaching the Earth: this would lead to a permanent change of the Earth's trajectory: dynamical systems have no way of forgeting perturbations.

This is no longer the case when we include dissipation. A damped pendulum will reach a position of equilibrium, whatever the initial perturbation.

Now, when we drive a system far from equilibrium, the 'attractor' which dominates the behaviour of the system near equilibrium may become unstable, as a result of the flow of matter and energy which we direct at the system. Non-equilibrium becomes a source of order; new types of attractors, more complicated ones, may appear, and give to the system a new space-time organization. Let us consider two examples which are widely studied today.

The so-called Bénard instability is a striking examples of instability giving rise to spontaneous self-organization; the instability is due to a vertical temperature gradient set up in a horizontal liquid layer. The lower face is maintained at a given temperature, higher than that of the upper. As a result of these boundary conditions, a permanent heat flux is set up, moving from bottom to top.

For small differences of temperature, heat can be conveyed by conduction, without any convection; but when the imposed temperature gradient reaches a threshold value, the stationary state (the fluid's state of 'rest') becomes unstable: convection arises, corresponding to the coherent motion of a huge number of molecules, increasing the rate of heat transfer. In appropriate conditions, the convection produces a complex spatial organization in the system.

There is another way of looking at this phenomenon. There are two elements involved: heat flow and gravitation. Under equilibrium conditions, the force of gravitation has hardly any effects on a thin layer of the order of 10 mm. In contrast, far from equilibrium, gravitation gives rise to macroscopic structures.

Non-equilibrium matter is much more sensitive to its environment than matter at equilibrium. I like to say that at equilibrium, matter is blind; far from equilibrium it may begin to 'see'.

Consider next chemical oscillations. We study a chemical reaction whose state we control through the appropriate injection of chemical products and the elimination of waste products. Suppose that two of the components are formed respectively by red and blue molecules in comparable quantities. We would expect to observe some kind of blurred colour with perhaps occasionally some flash of red or blue spots. This is, however, not what actually happens. For a whole class of such chemical reactions, we see in sequence the whole vessel become red, then blue, then red again: we have a 'chemical clock'. This violates our intuition about chemical reactions.[9]

We used to speak of chemical reactions as being produced by molecules moving in a disordered fashion and colliding at random. But, in order to synchronize their periodic change, the molecules must be able to 'communicate'. In other words, we are dealing here with new supermolecular scales—both in time and space—produced by chemical activity.

The basic conditions to be satisfied for such chemical oscillations to occur is auto- or cross-catalytic relations, leading to 'nonlinear' behaviour, such as described in numerous studies of modern biochemistry. Remember that nucleic acids produce proteins, which, in turn, lead to the formation of nucleic acids. There is an autocatalytic loop involving proteins and nucleic acids.

Nonlinearity and far-from equilibrium situations are closely related; their effect is that they lead to a multiplicity of stable states (in contrast to near-from-equilibrium situations, where we find only one stable state).

This multiplicity can be seen on a bifurcation diagram where we plot the solution of some nonlinear problem against a bifurcation parameter (for example, the concentration of some chemical component versus the time of sojourn of the molecules in a chemical reactor). For some critical value of this time, new solutions emerge. Moreover, near the bifurcation point, the system has a 'choice' between two branches: we therefore expect fluctuations to play an essential role.

We have mentioned the fact that dissipative systems may forget perturbations: these systems are characterized by attractors. The most elementary attractors are points or lines. But attractors may present a more complex structure; they may be formed of a set of points. Their distribution may be dense enough to permit us to ascribe to them a fractal dimensionality.[10]

Such systems have unique properties, reminiscent of the turbulence that we encounter every day. They combine both fluctuations and stability. The system is driven to the attractor, however, as this one is formed by so 'many' points, we may expect large fluctuations. One often speaks of 'attracting chaos'. These large fluctuations are connected to a great sensitivity with respect to the initial conditions. The distance between neighbouring trajectories grows exponentially in time. Attracting chaos has now been observed in a series of situations including chemical systems or hydrodynamics; but the importance of these new concepts goes far beyond proper physics and chemistry. We have already indicated the examples of long-term behaviour of climate or the electrical activity of the brain; there is no doubt that the new concepts are essential features of our environment; their study will permit us to model complex behaviour displayed by systems in ecology or economics.

Interface

The physics and chemistry of complex phenomena leads to a new interface between 'pure' and 'applied' research. This interface is growing so rapidly that I can give only a few brief examples.

A characteristic feature of far-from-equilibrium conditions is the possibility of bistability. For given boundary conditions, there may be more than one stable solution.[8]

A remarkable application of bistability is in optonics, in which the intensity of a coherent light beam through a resonant cavity may induce more than one stable value of the transmitted intensity.[11,12] This bistability appears as a transposition to optics of the hysteresis phenomenon well known in magnetism.

The stable states of the system are a function of its history, and not only of the boundary conditions: for a given value of the incident light intensity, it will evolve towards the low transmission branch (opaque state) if it enters the bistable zone coming from below, or towards the high transmission branch (transparent state) if it comes from above. It therefore acts as a binary memory.

Potential advantages of optical memories are: three orders of magnitude as far as speed of response is concerned (from 10^{-9} to 10^{-12} seconds); and parallel processing, as a bistable optical element whose section is 1 cm^2 may easily process in parallel more than a thousand pieces of information. What is perhaps more important, these components are likely to act as optical transistors.

It is interesting that this phenomenon of bistability is present in many problems, for example, in biological cell dynamics. A simple example, which is being studied by my colleagues in Brussels, is the interaction between tumour cells and immune system cells which kill tumour cells.[13] Most of the effort in studying cancer is directed to discovering the mechanisms which lead to the transformation of a cell to a cancerous cell. In contrast, here we concentrate on the response of the organism to a given population of cancer cells. Basically, a minimal dynamical model would be one in which cancer cells form complexes with cytotoxic cells, which are then regenerated after having killed cancer cells. This situation may lead to one or many steady states. One observes that each cytotoxic cell may bind more than one tumour cell; this leads to highly non-linear processes; for this reason, one has to expect multiple states. In this perspective, one of the main approaches of cancer would be to study the transition from a dormant form of cancer to a virulent one.

Other recent research concerns nucleation of fractures and the initiation of plasticity in materials submitted to stress.[14] As is well known, every material contains defects. Under stress, some immobile dislocation may become mobile and interact. There is then an obvious analogy with the reaction/diffusion equations, which have been widely studied for chemical systems that are far from equilibrium.

In conditions involving stress, there is a possibility of spatial dislocation patterns, leading to an accumulation of dislocations in some regions. These regions, which have been observed experimentally, are then likely to lead to fractures and plasticity.[15,16]

I will also mention two types of problems which, in addition to nonlinearity, involve fluctuations. Ideally speaking, for systems presenting a bifurcation point leading from one stable solution to two stable solutions, the probabilities of selecting one branch against the other are equal. But completely symmetrical solutions are only limiting cases. Currently, we are dealing with *imperfect* bifurcations, which can play a crucial role in the selection of the outcome. An extreme example is the selection of chiral molecules, in which a very small difference in the energy of formation of the molecule could lead to preferential selection. This is basically due to the possibility of polarizing the fluctuations near the bifurcation point.[17]

We can now begin to understand other cases, some of which have great potential importance, such as combustion and ignition where the deterministic description breaks down.[18] We have an initial induction regime, characterized by a very small rate of change which is followed by violent explosive behaviour. As the result of the induction stage, fluctuations play an important role: one finds a statistical distribution of ignition times instead of a simple, deterministic ignition time.

Abnormal fluctuations have also been observed in many biological problems, such as the distribution of growth rate of young males and females near puberty. This also indicates the existence of an autocatalytic effect, with a long induction period, as is the case in combustion. It would be fascinating to examine these ideas in learning processes, which often proceed by steps as indicated by Piaget, and are likely to have long inductive periods.

Irreversible processes

The discovery of the constructive role of irreversible processes in physics and chemistry, and of their importance in understanding physical processes as well as the behaviour of the biosphere, leads us to reconsider the microscopic meaning of irreversibility. Traditionally, irreversibility was only tolerated on the macroscopic level. It was supposed to be the result of ignorance of the exact dynamical state of the system. In contrast, on the basic microscopic level, there would be no question of an arrow of time, and no irreversibility.

This problem is closely related to the transition from the basic description involved in classical or quantum mechanics, with its deterministic and time-reversible features, to a description in which probability and irreversibility play a fundamental role. Only a few years ago, this problem seemed impossible to solve. The two descriptions, the dynamical one and the thermodynamical one, seemed to be separated by a gap which could not be bridged.

We begin now to see a way out of this difficulty. I will now briefly describe the basic ideas involved, as this is one of the most fascinating conceptual problems of contemporary physics. I shall limit myself to the case corresponding to highly unstable classical dynamical systems; but an extension of this approach to a more general class of classical systems[19,20] as well as to quantum mechanics[21] have also been worked out.

Classical mechanics may be described in the following two ways:

First, as the motion of a point ω in phase space Γ

$$\omega_t = S_t \, \omega_o \tag{1}$$

For Hamiltonian systems, and more generally for 'flows', the mapping S_t of Γ to itself is inevitable: the configuration ω_o defines all the past and future configurations ω_t defines the 'trajectory' of the system.

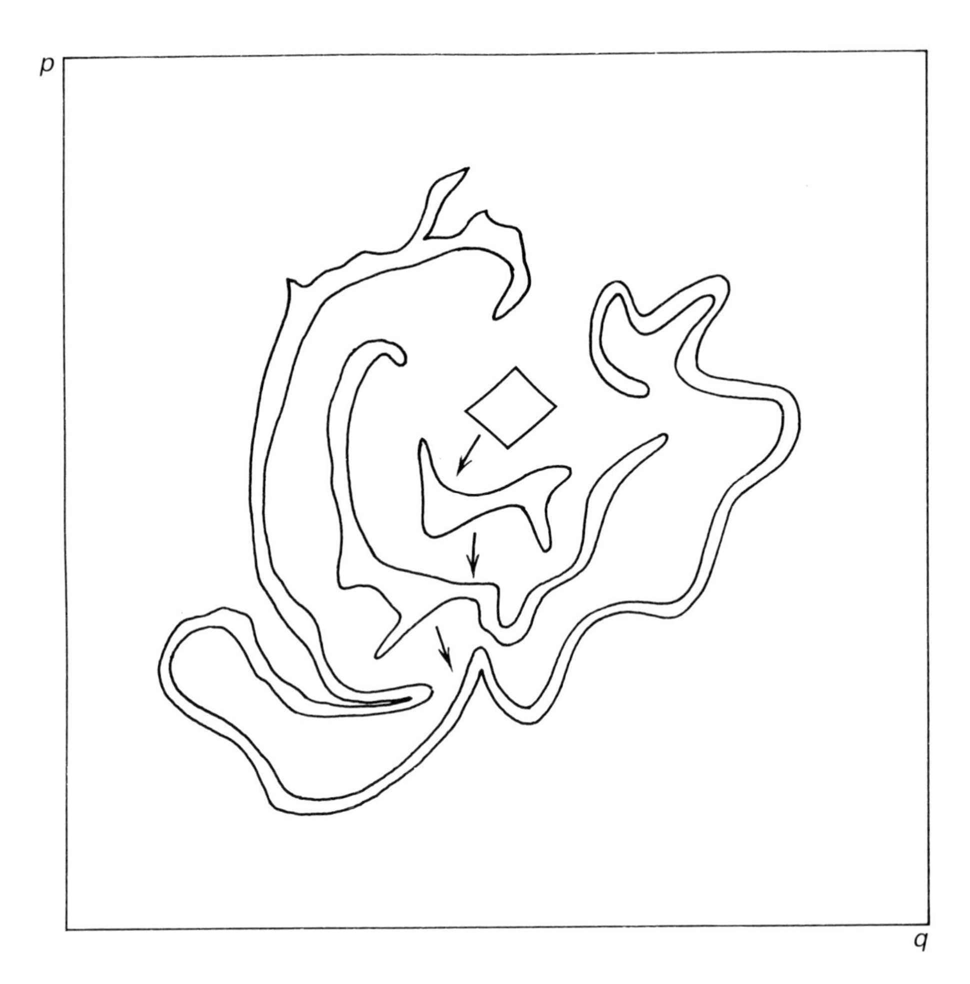

Figure 1 Deformation of a volume representing some initial distribution of states in phase space, where q refers to the position in original space R, and p to the momentum.

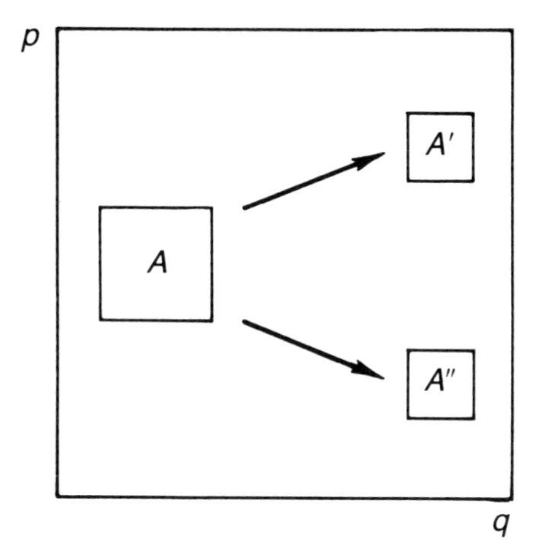

Figure 2 Fragmentation of volume in phase space.

Suppose we make sequential measurements (separated by some given time interval). The trajectory can then be visualized as an infinite sequence of numbers $\{\omega_n\}$ (with $-\infty \leqslant n \leqslant +\infty$).

This description expresses clearly the time-reversibility inherent to classical mechanics. The sequence $\{\omega_n\}$ exists for all time. The flow of time leads simply to a relabelling of the points ω_n.

Second, we may also consider a statistical description in terms of a Gibbs ensemble $\varrho(\omega)$. The evolution of the system is then described by the unitary operator U_t.

$$\varrho_t(\omega) = U_t\varrho(\omega) = \varrho_o(S_{-t}\omega) \tag{2}$$

A fundamental theorem in classical mechanics (the Liouville theorem) states that the volume in phase space (the 'measure') is preserved; however, as shown in Figure 1, this volume may be highly deformed, and even fragmented, as in Figure 2.

The deformation or fragmentation of the initial volume gives the 'appearance' of an approach to equilibrium, in which all the points would be uniformly distributed in the phase space Γ. But this is only an appearance, as the information (or entropy) is constant in time:

$$\int \varrho_t \log \varrho_t d\omega = \text{constant in time} \tag{3}$$

The mechanical cosmos has no history. In this description, the world would be similar to a museum in which information is stored forever. Time could then be compared with a tornado, which breaks a house into pieces. But the pieces remain, and could be used to reconstruct the house. For nearly three centuries after Newton, classical dynamics appeared as a closed system based on an axiomatic formulation. Recent achievements in dynamics have shown that this is not so.

We know that there exist classes of dynamical systems which are highly unstable (such as the Baker Transformation, to be introduced below). The sequence $\{\omega_n\}$ describing a trajectory corresponds to an infinite sequence of random numbers ω_n.[22]

Such an infinite sequence corresponds to an uncomputable algorithm. It is precisely for this class of unstable dynamical systems (technically called K-flows, from the Russian mathematician Kolmogoroff) that we can now go rigorously from dynamics to thermodynamics, or more precisely from the unitary evolution as described by U_t (see equation (2)) to a Markoff process W_t through an appropriate 'deformation' of the Gibbs ensemble $\hat{\varrho}$ mediated by an appropriate linear operator Λ:

$$\hat{\varrho} = \Lambda \varrho \tag{4}$$

such that

$$\hat{\varrho}_t = W_t \, \hat{\varrho}_o \tag{5}$$

where W_t corresponds to a Markoff process. We may then associate with the entropy

$$\int \varrho_t \log \varrho_t d\omega \tag{6}$$

which varies monotonously in time.

In contrast with dynamics, Markoff processes correspond to a *contracting* semigroup, in which information is no longer 'stored' as in dynamics, but damped and ultimately destroyed, as time passes.

Relations (2), (4) and (5) imply that

$$W_t = \Lambda U_t \Lambda^{-1} \tag{7}$$

There exists, therefore, a simple relation between the dynamical and the thermodynamical description. However, in this perspective, dynamics corresponds to an idealization, while the 'real' world, which includes the second law of thermodynamics, is described by W_t. This idealization remains fundamental, as it is used as the starting point for the construction of W_t.

We cannot, of course, go into the details of the construction of Λ;[23] so let us make a few qualitative remarks. For unstable dynamical systems (the K-flows), the construction of Λ is most conveniently given in terms of the operator T of 'internal time'. This is, by definition, a self-adjoint operator T on $h_{eq}^{\perp}$ (the subspace orthogonal to constant functions) satisfying:

$$U^{-n} T U^{n} = T + nI \tag{8}$$

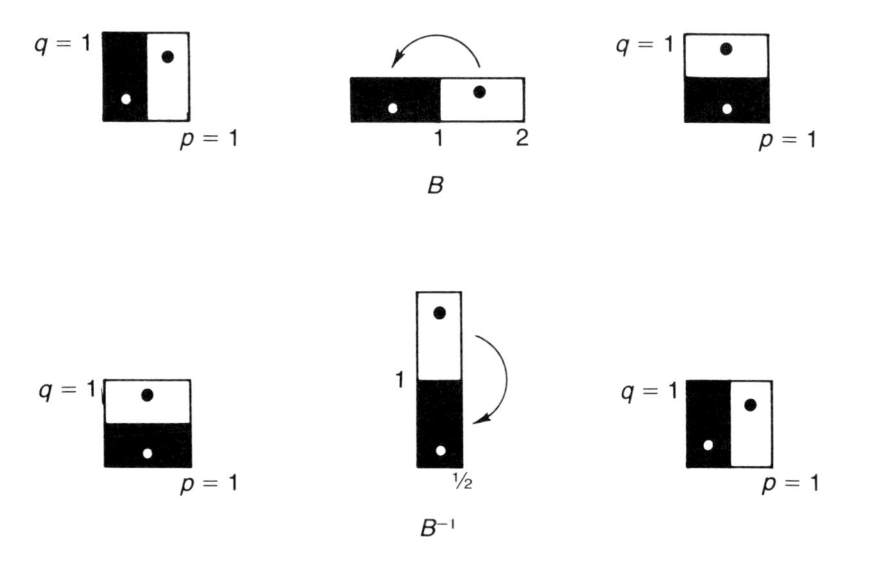

Figure 3 Realization of the Baker Transformation (B) and its inverse (B^{-1}).

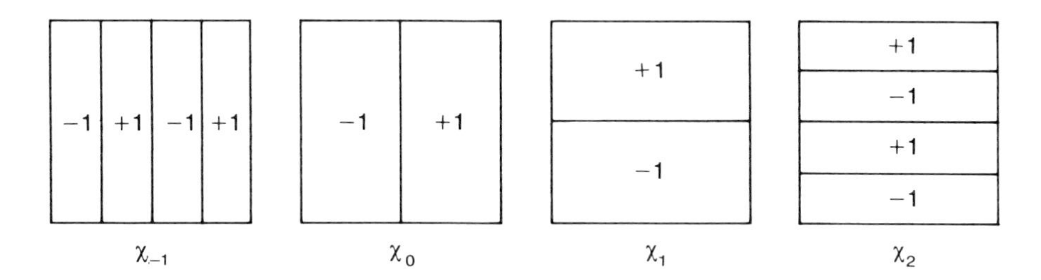

Figure 4 Some functions χ_n generated from χ_o through the iterated action of the operator U.

where we use the index n (integer) to emphasize that we consider discrete time intervals. To grasp the idea of an internal time, we may look at some dynamical systems, such as the so-called 'Baker Transformation'. Here the phase space is a square; the transformation coresponds to the well-known operations of the Baker: flatten, cut, fold, as represented on Figure 3. The sequence of states for a given point is clearly deterministic. But any region, whatever its size, contains trajectories diverging at each fragmentation.

A complete set of orthogonal eigenfunctions of T can be constructed as follows: let χ_o be the function which assumes the value -1 on the left half of the square and $+1$ on the right half. define $\chi_m = U^n\chi_o$. A few of these functions are represented in Figure 4.

A complete set of eigenfunctions of T is obtained by taking all possible finite products of χ_m. Such a product belongs to the eigenvalue n, when n is the maximum of the indices m of χ_m appearing in the product. For example, $\chi_{-5}\,\chi_3$ and χ_3 are both eigenvectors of T corresponding to eigenvalue $+3$. We shall denote by $\varphi_{n,i}$ a complete set of eigenvectors of T; the index n labels the eigenvalue of T and the index i labels the additional degeneracy. The eigenfunctions $\varphi_{n,i}$ together with the constant function 1 form a complete set of orthogonal functions. Let E_n denote the subspace spanned by all the eigenvectors of T corresponding to eigenvalue n. Then the operator Λ may be expressed as:

$$\Lambda = \sum_{n=-\infty}^{+\infty} \lambda_n E_n + |1\rangle\langle 1| \tag{9}$$

where $|1\rangle\langle 1|$ denotes the projection on the constant function 1, and λ_n is a suitable nonincreasing sequence of numbers. To understand the physical meaning of Λ it is important to introduce the concept of 'contracting' and 'dilating' fibres. These are peculiar regions in phase space, whose fate are contrasting. Let us suppose that a distribution of states is concentrated on a vertical line: at each Baker Transformation, this line is reduced, and would reduce to a point in the far distant future. In contrast, a distribution corresponding to a horizontal line will be duplicated, and would virtually occupy the entire phase space. Clearly, these evolutions correspond to two opposite dynamics, each involving symmetry breaking in time: dilating fibres correspond to equilibrium in the distant future, while contracting fibres correspond to equilibrium in the distant past.

We can now formulate the main properties of Λ:

(1) Λ is a *non-local* operator; if we apply it to a point (represented by a δ-function $\delta(\omega-\omega_o)$, we generate an *ensemble* of points.

$$\Lambda\ \delta(\omega-\omega_o) \rightarrow \text{ensemble of points.}$$

In a simple case studied in ref. 23, $\Lambda\ \delta(\omega-\omega_o)$ leads to the contracting fiber going through the point ω_o. The non-locality is, of course, the reason why we go from a deterministic description to a probabilistic one. The introduction of probability is not due to our ignorance, but to the dynamical instability inherent to these systems, and permits the construction of Λ in terms of dynamical concepts.

(2) Λ breaks the temporal symmetry. Only distributions $\hat{\varrho}$ leading to a finite value of the 'entropy' $\int \hat{\varrho} \log \hat{\varrho}\, \delta\omega$ are admissible: this excludes points as well as contracting fibres.[22] As a result, only distributions which lead to equilibrium in the future can be observed experimentally or be prepared. We see that the distributions $\hat{\varrho}$ describing the microscopic level already presents the basic features of the macroscopic world: temporal polarization and uncertainty of the future.

We begin, therefore, to decipher the message embedded in the second law of thermodynamics: the world is made up of unstable dynamical systems which are temporally polarized.

For unstable systems, which have a privileged time direction, we cannot impose initial conditions which would force an ensemble of points to concentrate on a single point. The future remains open.

The message carried by the second law is, therefore, not one of ignorance and subjectivity. On the contrary, it gives us some fundamental information about the overall structure of the physical world.

Initially, we referred to a basic level of physical description. We now have to take into account the second law of thermodynamics, even on this level. Therefore, this level can be formed neither by trajectories nor by wave functions, which satisfy deterministic equations in which the *future* would be already included in the *present*. The basic level of physics would be formed by nonequilibrium ensembles, which are less well determined than trajectories or wave functions, and which evolve in the future in such a way as to increase this lack of determination.

If the world were built like the image designed for reversible, eternal systems by Galileo Galilei and Isaac Newton, there would be no place for irreversible phenomena such as chemical reactions or biological processes.

We have seen that for unstable dynamical systems, a trajectory can be viewed as a double infinite sequence of random numbers. We have also mentioned that this is excluded as it could require an infinite amount of information measured by $\int \hat{\varrho} \log \hat{\varrho} \delta\omega$.

In contrast, we can only reach a finite sequence, which would give a finite value to this information. In this sense, Our knowledge of the world corresponds to a 'temporal window': of the infinite span of eternity we may only grasp what appears in this window. In this view, God is no more an archivist unfolding an infinite sequence he had designed once and forever. He continues the labour of creation throughout time.

Conclusion

The universe has a history. This history includes the creation of complexity through mechanisms of bifurcation. These mechanisms act far from the equilibrium conditions as realized in the Earth's biosphere. They may also have been of special relevance to the early stage of the universe, where we have to expect a strong coupling between matter and gravitation.[24]

Non-equilibrium physics is a subject in a state of explosive growth. I have tried to show some of my fascination in this topic. It leads both to new applications of direct scientific and technical importance, and to new perspectives on the foundations of physics.

Rationality can no longer be identified with 'certainty', nor probability with ignorance, as has been the case in classical science. At all levels, in physics, in biology,[25] in human behaviour,[26] probability and irreversibility play an essential role. We are witnessing a new convergence between two 'visions of the world', the one emerging out of scientific experience, and the other we get from our personal existence, be it through introspection or through personal experience.

Sigmund Freud told us that the history of science is the history of an alienation: since Copernicus we no longer live at the centre of the universe; since Darwin, man is no longer different from other animals; and since Freud himself conscience is just the emerged part of a complex reality hidden from us.

Curiously, we now have the opposite view. With the role of duration and freedom so prevalent in human life, human existence appears to be the most striking realization of the basic laws of nature.

References

1. Prigogine, I., and Stengers, I., *La nouvelle Alliance*, Gallimard Paris (1979); Engl. transl., *Order out of Chaos*, Heinemann, London (1984).
2. Prigogine, I., *From Being to Becoming*, Freeman, San Francisco (1979).
3. Nicolis, G., and Prigogine, I., *Exploring Complexity*, Piper Vlg, Munich (in press).
4. Nicolis, C., and Nicolis, G., *Nature*, *311*, 529–532 (1983).
5. Babloyantz, A., and Nicolis, C., *J. Theor. Biol.*, (submitted).
6. Babloyantz, A., Salazar, J.M., and Nicolis, C., *Phys. Lett.* (submitted).
7. Petrosky, T.Y., "Chaos and irreversibility in a conservative nonlinear dynamical system with a few degrees of freedom", *Phys. Rev.*, *29*, (4) 2078–2091 (1984).
8. Nicolis, G., and Prigogine, I., *Self-Organization in Nonequilibrium Systems*, Wiley, New York (1977).
9. Field, R., and Burge, M., (Editors), *Oscillations and Travelling Waves in Chemical Systems*, Wiley, New York (1985).
10. Tomita, K., "The Significance of the Concept 'Chaos'", *Prog. Theor. Phys.*, Suppl. No. 79, 1–25 (1984).
11. Abraham, E., and Smith, S.D., "Optical Stability and Related Devices", *Rep. Prog. Phys.*, *45*, 815–885 (1982).
12. Lugiato, L.A., "Theory of Optical Bistability", *Prog. Opt.*, *21*, 71–216 (1984).
13. Prigogine, I., and Lefever, R., "Stability Problems in Cancer Growth and Nucleation", *Comp. Bioch. Physiol.*, *67B*, 389 (1980).
14. Walgraef, D., and Aifantis, E., "Dislocation patterning in fatiguated metals as a result of dynamical instabilities", *J. Appl. Phys.*, (in press).
15. Mugrabi, H. in Brulin, O. and Hsieh, R.K.T. (Editors), *Continuous Models of Discrete Systems*, 4, pp. 241–257, Amsterdam, North Holland (1981).
16. Tabata, T., Fujita, H., Hiaraoka, M. and Onishi, K., *Philos. Mag.*, *A 47*, 841 (1983).
17. Kondepudi, D.K. and Nelson, G.W., *Phys. Rev. Lett.* *50*, (14), 1023–1026 (1983).

18. Nicolis, G., Baras, F., and Malek-Mansour, M., "Stochastic Aspects of Nonequilibrium Transitions in Chemical Systems", in Vidal, C. and Pacault, A. (Editors), *Nonequilibrium Dynamics in Chemical Systems*, Springer, Berlin (1984).
19. Petrovsky, T., and Prigogine, I., "Chaos and nonunitary evolution in nonintegrable Hamiltonian systems", *Phys. Rev. A*, (submitted).
20. Petrovsky, T., "Chaos and irreversibility in a conservative nonlinear dynamical system with a few degrees of liberty", *Phys. Rev. A, 29* (4) (1984).
21. George, Cl., Mayne, F., and Prigogine, I., "Scattering Theory in Superspace", *Advances in Chemical Physics.*
22. Ford, J., "How Random is a Coin Toss?", in Horton Jr, C.W., Reichl, L., and Szebehely, V.G. (Editors), *Long-Time prediction in Dynamics*, Wiley, New York (1983).
23. Misra, B., and Prigogine, I., "Irreversibility and Nonlocality", *Lett. Math. Phys., 7*, 421–429 (1983).
24. Géhéniau, J., and Prigogine, I., *PNAS* (submitted).
25. Deneubourg, J.L., Pasteels, J.M., and Verhaeghe, J.C., "Probabilistic Behaviour in Ants: A Strategy of Errors?", *J. Theor. Biol. 105*, 259–271 (1983).
26. Allen, P.M., Engelen, G., and Sanglier, M., "New Methods for Policy Exploration in Complex Systems", *Spec. Issue Envir. Plann., B 12* (1) 1–138 (1985).

Professor Paul Davies

Paul Davies is Professor of Theoretical Physics at the University of Newcastle-upon-Tyne.

He graduated in Physics, and followed with PhD in Theoretical Physics in 1970. After two years as Visiting Fellow at the Institute of Theoretical Astronomy in Cambridge, he became Lecturer in Applied Mathematics at King's College, University of London, and took up his present appointment in 1980.

He is active in research across a wide field of fundamental physics and cosmology. Special interests are the origin and end of the universe, high-energy processes in the big bang, the nature of time and the origin of time asymmetry, quantum field theory, quantum aspects of gravity, particle creation in strong gravitational fields, black holes and their thermodynamic properties, the physics of extra dimensions, and the conceptual basis of quantum mechanics.

His work is reflected in over sixty technical research papers in specialist journals, several hundred popular and review articles, and eleven books that have collectively been translated into eight non-English languages.

He lectures widely at both popular and specialist level, and contributes regularly to newspapers, magazines, journals, and science documentaries for radio and television.

The unfolding universe

The discovery of the first dinosaur fossil occurred by accident in 1822 when Mrs Mary Mantell noticed among a pile of quarried stones, a piece of sandstone containing several large teeth. Her husband who was an amateur geologist, made the bold conjecture that the creature had inhabited the Earth before mammals existed. He identified the creature as *Iguanadon*.

This accidental discovery came at a critical time for science. The age of the Earth had traditionally been thought to be several thousand years, however, by the end of the eighteenth century most geologists had begun to recognize that a much longer period of time was necessary for the completion of activities such as sedimentation and erosion. By the middle of the nineteenth century the time-span had been stretched to thousands of millions of years. Today, the Earth has been radioactively dated at 4600 million years.

This dinosaur fossil soon came to be recognized as the remains of extinct creatures that roamed the Earth between about 200 and 65 million years ago. Subsequent, more diligent searches have uncovered fossil remains of living organisms that date from at least 3000 million years and possibly nearly 4000 million years.

Although most people associate fossils with the frozen imprints of once-living creatures, there are many other physical objects that are imprinted with an inanimate record of the remote past. For example, the pock-marked surface of the Moon, Mars, and Mercury bear witness to a phase of violent bombardment at the dawn of the solar system. In a sense, all physical things are fossils. Every object that exists possesses some sort of history and encodes information about the circumstances that brought it into being. The trick is to be smart enough to decode the information. For example, we can examine our bodies and ask what they can tell us about the past.

Biological information is encoded in our genes which are built from molecules of DNA. All life on Earth is based on DNA which may, therefore, be regarded as a relic of the origin of Earth life. Our particular genetic structure bears countless imprints of the physical conditions encountered by our ancestors, and which helped to shape the evolutionary pathway followed by our species. Our bodies are thus living fossils that embody a coded history of our planet.

Biological information relates to the way in which the atoms of carbon, hydrogen, oxygen, and other elements within living organisms are strung together into complex forms. But what about the atoms themselves, the raw material from which our bodies and all objects around us are made?

According to modern ideas of cosmology, these atoms are relics of physical processes that occurred long ago, out in the depths of the universe. They are cosmic fossils. The primary constituent of cosmic material is hydrogen, with helium accounting for about 10 per cent of all atoms, and the 90 or so other elements representing only a minute fraction of the whole. Much of the material inside us, then, consists of cosmic trace elements, hugely concentrated. Their origin must be sought in the complex processes that occur inside stars.

When the universe began, the cosmic material contained essentially no medium or heavy nuclei. These elements are the ashes of the nuclear fires that sustain the stars. In a star like the Sun, the core is a nuclear fusion reactor, in which the fuel consists primarily of hydrogen nuclei (protons). The intense heat of the solar furnace agitates the protons with such violence that occasional very close encounters occur, even though protons will repel each other with a powerful electric force. Should the colliding protons come within range of the strong nuclear force, fusion is possible. A nucleus consisting of two protons is unstable, but if one proton transmutes into a neutron through a weak interaction (essentially the inverse of beta decay) then a stable nucleus of deuterium forms, with the release of energy, which helps keep the furnace hot. Further fusion reactions result in the conversion of deuterium into helium. In old stars the synthesis of heavier nuclei out of light ones is more developed. Successive fusion processes produce first carbon and then a whole sequence of ever-more complex nuclei.

As a star approaches the limit of its fuel reserves, its internal structure resembles the skins of an onion with layer upon layer of different chemical elements representing various stages in the long sequence of synthesis. Over its lifetime, an old star has gradually been transformed from almost pure primordial hydrogen and helium into a repository for spent nuclear ash, in the form of heavy chemical elements. During its final phase of activity, such a star may become unstable. The faltering nuclear reactions are unable to sustain the intense internal heat and pressure needed to support the star against its own tremendous weight. Gravity then runs out of control, imploding the star's core in an instant. A huge pulse of energy in the form of neutrinos and shock waves released from the core blasts the outer layers of the star into space, spewing the heavy elements into the depths of the galaxy. This outburst is called a supernova explosion. Each one enriches the galactic material with trace elements so vital in the formation of solid planets and the life-forms that inhabit them. Our bodies are, therefore, built from the fossilized debris of once-bright stars that annihilated themselves aeons before the Earth or Sun existed.

The heavy elements about us record a violent history, but the light elements, hydrogen and helium, date from a still more violent epoch in our cosmic history—the big bang. But were these elements there from the beginning or are they fossils from some very early phase?

The intense heat which accompanied the big bang provides the key to an understanding of the early universe. In its simplest form, the hot big bang theory assumes that the universe exploded spontaneously into existence from a state of infinite compression and infinite heat. As the expansion proceeded, so the temperature fell from infinity, rapidly at first, and then more slowly, until the universe had cooled enough for stars and galaxies to form, only then did the first atoms form. Atoms, then, are relics from 100 000 years after the creation.

What is the origin of the nuclei of hydrogen and helium? During the first few minutes after the big bang, the temperature of the cosmic plasma was in excess of 10^6 K, which is hot enough for nuclear reactions to occur. Using computer models combined with nuclear data, astrophysicists can reconstruct the details of the nuclear activity which took place in the first minutes of the universe.

At the end of the first second the temperature was 10^{10} K—too hot for composite nuclei to have existed. Instead, a 'soup' of individual protons and neutrons in chaotic motion filled all of space, mingled with electrons, neutrinos and photons (heat radiation). The early universe expanded extremely rapidly, so that when one minute had elapsed the temperature had dropped to a 10^8 K, and after several minutes it had fallen below the level at which nuclear reactions are possible. There was thus a relatively brief period of a few minutes during which the protons and neutrons could aggregate together into composite nuclei.

The principal nuclear reaction was the fusion of protons and neutrons to form helium nuclei. Because protons are marginally lighter than neutrons, they existed in greater abundance, so that when the production of helium was complete, some protons remained free. Calculations show that very little else happened in the short time available; thus the composition of the emerging plasma was about 10 per cent helium nuclei and 90 per cent hydrogen nuclei, which reflects with satisfying precision the observed abundances of these elements in the universe today. The conclusion is that the element helium is a fossil remnant from the primeval furnace which raged during the first few minutes following the creation.

It is fortunate that the primeval material was over-rich in protons, for it is from the residue of unmatched protons that the universe derives its hydrogen. Without hydrogen the Sun would not burn, nor would there be any water in the cosmos. It seems unlikely that life could exist in these circumstances.

Fossils from the first second

To make sense of the events that unfolded in the very early universe, it is necessary to understand the nature of cosmic activity. Throughout the Earth's 4600 million year history change has been slow. Geological time-scales are reckoned in millions of years. If we could journey back to a few million rather than a few thousand million years from the big bang, we should find things happening much faster. Galaxies were being formed in the space of a few hundred million years, while stars formed even more quickly, perhaps in a few tens of millions.

Back before 100000 years the universe was relatively featureless. This is the phase of glowing plasma. The pace of events can be gauged by the rate of cosmic expansion and the rate at which the temperature was falling. The expansion was some 100000 times faster then than it is today. The temperature was several thousand degrees. At earlier times the rate was faster still and the temperature higher. At time 1 s the universe was doubling in size in about a second, its temperature 10^{10} K. Within the first second the pace of change escalated still faster, rising without limit as the instant of creation was approached.

Mathematically, this accelerating rate of activity is described as a 'reciprocal' relationship. For example, the expansion rate is proportional to $1/t$ and the temperature to $1/\sqrt{t}$, where t is the time since the creation. As t becomes smaller and smaller, so these quantities rise faster and faster towards infinite values. Because the level of activity rises steeply as we journey back in time to the first moment, important changes are likely to occur in briefer and briefer time intervals. It is then more meaningful to adopt a 'power of ten' approach to time. For example, as much happened during the interval from 0.1 s up to 1 s as occurred in the interval from 0.01 s up to 0.1 s. And so on. Each time the interval is further subdivided by ten so we encounter a comparable degree of change compressed into that smaller interval.

How far back can we extrapolate our model of the early universe with any sort of confidence. It often comes as a surprise to learn that the extreme conditions which prevailed during most of the first second of the universe are within experimental experience. Modern particle accelerators can, for a brief instant, simulate the physical conditions that occurred as early as 10^{-12} s, when the temperature was a staggering 10^{16} K and today's entire observable universe was compressed into a region no larger than the solar system.

The farther back we probe, then, the more extreme physical conditions we encounter. The most important parameter to gauge our progress is energy. The energy possessed by a typical particle in the primeval 'soup' or plasma rises ever more sharply as the first moment is approached. At 1 min we have to deal with nuclear energies. At 1 s we reach energies that can be attained by some radioactive emissions. At 1 microsecond (1 μs), the energy of a typical particle is comparable to that in the early particle accelerators. When we reach 10^{-12} s (1 picosecond [1 ps]), we are approaching the frontiers of current high-energy particle physics. Beyond this point, theory is our only aid.

As the energy is raised, so forces begin to merge. First the electromagnetic force merges with the weak force. This occurs at an energy equivalent to about 90 proton masses, corresponding to a temperature of around 10^{15} K. Existing accelerators can just reach this regime, which is where W and Z particles are liberated. The further merging of the electroweak and strong forces, and eventually gravity, does not occur until enormously greater energies have been reached. We have to attain the unification and Planck scales, trillions of times more energetic than the electroweak scale.

At 10^{-12} s after the big bang the temperature was so high that all the now familiar particles and antiparticles would have been created out of the available heat energy. The universe contained almost equal proportions of matter and antimatter. Later, when the particle–antiparticle pairs which constituted the majority of material annihilated, a residue of matter was left. The density of particles was so high that an equilibrium was reached, in which the available energy was partitioned democratically among all the different particle species.

The nature of the cosmic material at this stage was unlike anything of which we have direct experience. Being packed together so densely, the hadrons did not possess individual identities. Protons and neutrons did not exist as separate entities. Instead, the cosmic material consisted of a fluid of quarks, moving about more or less independently. Furthermore, at these energies all distinction between the weak and the electromagnetic force was lost, and the nature of the leptons and quarks was peculiar indeed. Particles such as electrons, muons, and neutrinos that we see today did not exist in their familiar forms. Photons, Ws, and Zs were inextricably conflated in identity.

The key to understanding this weird high-temperature phase of matter is symmetry-breaking. A gauge symmetry can be spontaneously broken to supply particles with masses and provide the distinction between the electromagnetic and weak forces. There is a general rule in nature that high temperatures tend to restore symmetry.

Our picture of the universe at 1 ps (10^{-12} s) is thus a remarkable one. The universe is filled with a mysterious fluid medium, unknown anywhere in the cosmos since. But this peculiar phase of matter cannot persist. As the temperature drops, a sudden phase change occurs. Abruptly, all the familiar particles—electrons, neutrinos, photons, quarks—are identifiable. The gauge symmetry has been broken, and the electromagnetic force separates out from the weak force.

Forward in time, another crucial phase change occurs at about 1 ms (10^{-3}s). The tight press of agitated quarks suddenly congeals into a sea of well-defined hadrons. Individual protons, neutrons, mesons, and other strongly interacting particles can now be discerned, with the quarks linked together in distinct units of two and three. Later as the temperature sinks yet lower, all the remaining antiparticles, such as the positrons annihilate, producing large amounts of gamma radiation, and the cosmic material now contains the more familiar mix of protons, neutrons, electrons, neutrinos, and photons that set the stage for helium synthesis after a few seconds elapse.

We can now see that protons and neutrons—the building blocks of the universe—did not always exist, but congealed from a broth of quarks at about 10^{-3} s. These nuclear particles can therefore be regarded as fossils of the first millisecond. Still more bizarre is the fact that the leptons and quarks which go to make up all matter, only achieved their present identities at about 10^{-12}. They are fossils from the first picosecond.

We can trace the origin of the elements back to distant epochs of star burning, and nucleosynthesis in the first minutes of the universe. The protons and neutrons that go to build these elements find their origin at still earlier moments, while the leptons, and the quarks that in turn build the nuclear particles, are relics of a time when the universe had existed for a mere one-million-millionth of a second. But a mystery still remains, a mystery that takes us back to an epoch much earlier still—to the so-called GUT (Grand Unified Theory) era.

The origin of matter

When the big bang theory was first proposed, cosmologists had to fall back on the assumption that the matter from which the universe is constructed was present in the beginning. Today, the new cosmology provides a plausible explanation for the origin of matter, based on the activities of the superforce.

The possibility that matter may be created out of concentrated energy has long been known. And there was no lack of energy in the big bang to generate all the matter in the visible universe—about 10^{50} tonnes in total. But how did all this matter come to exist without an equal quantity of antimatter? When matter is created in the laboratory, antimatter is always produced and the symmetry between matter and antimatter seems to be deep-rooted in the laws of physics. But where has all the antimatter gone?

First, we have to be sure that the universe really is made entirely of matter. A rock made of antimatter would look in all respects like a rock made of matter. However, if each of them is brought into contact with a piece of matter, the antimatter rock will vanish amid an explosive outburst of nuclear proportions. Clearly, the Earth is made of 100 per cent matter.

But is this asymmetry true of the universe as a whole? As far as we can tell it is. If our galaxy contained substantial quantities of antimatter, the inevitable collisions that occur between gas, dust, stars, planets, and other objects would produce a deluge of gamma radiation as the antimatter encountered matter and annihilated itself. Gamma radiation at this intense level would certainly be detected, and astronomers have placed a limit on the antimatter content of our galaxy of one part in 10^3. Apart from the occasional antiprotons found in cosmic rays, the galaxy seems to be pretty well pure matter.

Ten years ago the only explanation offered for the primordial imbalance between matter and antimatter was to suppose that it was built in from the beginning, that the material issuing from the big bang had a disproportionate amount of matter as compared to antimatter.

Rather than assuming that the excess of matter in the universe is God-given, a more satisfying explanation is to suppose that initially there was complete symmetry between matter and antimatter and that a preponderance of matter developed *after* the beginning by natural causes, and then became 'frozen in' to the universe. It would then no longer be necessary to believe in an arbitrary initial condition; the state of exact equality (zero excess) is unique. The observed excess of matter over antimatter might then be explained quantitatively in terms of a physical theory.

For this idea to work a physical mechanism is needed that breaks the matter–antimatter symmetry. In the late 1970s, just such a mechanism came to hand in the guise of the grand unified theories (GUTs). One of the more sensational predictions of GUTs is that protons are unstable, and decay into positrons. The relationship between proton decay and matter–antimatter asymmetry can be seen by considering the possible long-term fate of a hydrogen atom (a proton plus an electron). When the proton decays, it emits a pion and a positron. The pion decays into two photons, while the positron can annihilate the electron to give two more photons. What started out as an atom of matter ends up as pure radiation energy. By this process, matter has been completely converted into energy, without encountering antimatter. Now every physical process can be reversed, which means that it is possible to conceive of energy turning into matter without the production of antimatter. This process, greatly speeded up, could explain how matter came to exist.

To model the creation process in detail, it is necessary to go back to the so-called GUT era. This means attempting to describe a universe only

10^{-32} s old! At that instant the cosmos would be filled with a broth of weird, unrecognizable particles, some of them extraordinarily heavy, packed to a density of 10^{73} kg m^{-3}, and bathed in heat at a temperature of 10^{28} K. The universe was then so young that light could not have made its way the equivalent of one-thousand-millionth of the distance across a proton since the first moment.

The crucial ingredients of this exotic broth are the superheavy particles that transmit the grand unified force, the so-called X particles which can introduce the lopsidedness between matter and antimatter. This is how: when an X decays it may yield many daughter particles of which, say, ⅔ may be matter but only ⅓ antimatter. The precise details of this asymmetry depend on the particular GUT adopted.

The primordial broth will also contain the antiparticles of X, usually denoted $\bar{X}$. If the universe started out symmetrically, there would have been equal proportions of X and $\bar{X}$. When the $\bar{X}$s decay, they reverse the asymmetry, yielding ⅔ antimatter to ⅓ matter. The net effect is that the initial symmetry remains intact.

To escape from this impasse, theorists assume that there must be a fundamental imbalance in the decay rates of X and $\bar{X}$. As a result, the $\bar{X}$ decays do not quite offset the X decays. There is perhaps a bias of about one in a thousand million in favour of X, giving us a one-in-a-thousand-million preponderance of matter over antimatter.

How reasonable is this assumption? In 1956, T.D. Lee and C.N. Yang, upturned the apple-cart by insisting that the weak force violates mirror symmetry. Until then physicists had assumed, that the forces of nature were indifferent to the distinction between left and right. There are many examples of natural structures with an inbuilt 'handedness', the most famous being DNA. Molecules of DNA have a shape like a spiral staircase winding always to the right. Though no left-handed DNA occurs naturally, there is no fundamental law of physics that prevents this. The fact that all Earth life is made from the right-handed form is presumably because the first self-replicating molecule just happened to be twisted that way. It is a good example of spontaneous symmetry-breaking; the actual structure is asymmetric even though the underlying physical forces remain symmetric.

When a physicist says that the forces of nature are mirror-symmetric he means that the fundamental processes induced by the forces would look equally permissible in a mirror as they do when viewed directly.

So confident were physicists that subatomic particles could not tell left from right that they did not test the matter. Then along came Lee and Yang who challenged the assumption. An experiment was quickly performed by Mrs C.S. Wu, who discovered that Lee and Yang were right. The weak force does violate mirror-symmetry. The Wu experiment, which involved measuring the numbers of left- and right-moving electrons emitted by carefully aligned radioactive cobalt nuclei, proved to be a turning point in physics. After that, no symmetries were safe.

In 1964, much interest was being taken in the behaviour of a particle called the neutral K meson. The idea that mirror symmetry was violated had by then been accepted, but it was assumed that antiparticles would always violate mirror symmetry in the reverse sense to particles. (Antiparticles usually display the reverse properties of particles.) If this were the case, there would be no way that the universe could generate a preponderance of matter over antimatter in the big bang, because for any process in which a particle was created there would be another mirror process somewhere in which an antiparticle was created. The peculiarities of the neutral K meson, which is a sort of hybrid particle–antiparticle, made it possible to test these ideas.

V.L. Fitch and J.W. Cronin found that mirror symmetry is *not* violated in equal and opposite ways by particles and antiparticles, at least as far as the K meson is concerned. There is a tiny, but highly significant lopsidedness here too. Such an asymmetry reflects a fundamental imbalance in the nature of the forces that drive some particle decays, providing concrete experimental evidence for asymmetry between matter and antimatter.

In the late 1970s, theorists began modelling the GUT phase of the big bang on the assumption that asymmetry really does exist in the grand unified force, and came up with numbers that suggest, typically, an imbalance between matter and antimatter of one in a thousand million. This means that for every thousand million antiparticles, a thousand million plus one particles are created. Although only tiny, the slight excess of particles proves to be absolutely crucial. When the universe eventually cools, the antimatter annihilates, and in so doing it destroys nearly all the matter. But one part in a thousand million excess of matter over antimatter remains left over. It is from this minute residue that all the objects in the universe, including man, are made. So ultimately, *all* matter is a fossil, a relic from the GUT era, a mere 10^{-32} s from the creation event.

If this analysis is to be believed, most matter that emerged from the big bang disappeared before the first few seconds had elapsed, along with all the cosmic antimatter. But this vanished material has left an echo of its erstwhile existence in the form of energy. The matter–antimatter annihilation produced about a thousand million gamma-ray photons for each electron and each proton that remained unscathed. Today this radiation has been cooled by the cosmic expansion, and forms the background heat radiation that fills the universe. Apart from the energy locked up in matter, this background heat accounts for the greater part of the energy of the universe. We thus have at hand a theory that not only explains how matter came to exist, but can also account for the ratio of matter to energy in the universe.

Before the event of GUTs, the temperature of the cosmic background heat radiation could not be explained. The level of radiant heat energy was another apparently arbitrary cosmic parameter, built into the universe at its creation. No reason was known why the temperature today could not be 0.3 or 30 K, rather than the 3 K it is. The GUTs provide a means to explain that temperature from physics. A present temperature of 3 K corresponds to about 10^9 photons to every proton and electron in the universe, and this value is in good agreement with the typical one-in-a-thousand-million excess of particles over antiparticles predicted by GUTs. One of the fundamental parameters of cosmology can thus be explained in terms of physical processes which occurred during the GUT era. It was at that unimaginably early moment of existence that the foundations were laid for the structure of the universe we see today.

The genesis paradox

What caused the big bang? This question is actually two rolled into one. We should like to know what triggered this explosive outburst in the first place. But if the big bang represents the origin of physical existence, including that of space and time, in what sense can anything be said to have *caused* this event?

On a purely physical level, the abrupt appearance of the universe in a huge explosion is a paradox. Of the four forces of nature which control the world, only gravity acts systematically on a cosmic scale, and in all our experience gravity is attractive. It is a pulling force. But the explosion which marked the creation of the universe would seem to require a pushing force of unimaginable power to blast the cosmos asunder and set it on a path of expansion which continues to this day.

People are often puzzledin the belief that if the universe is dominated by the force of gravity it ought to be contracting, not expanding. As a pulling force, gravity causes objects to implode rather than explode. In the very early universe, the compression of material exceeded that of even the densest star, and this fact often prompts the question of why the primeval cosmos did not itself turn into a black hole at the outset.

The traditional response leaves a credibility gap. It is argued that the primeval explosion must simply be accepted as an initial condition. Certainly, under the influence of gravity, the rate of cosmic expansion has continually slowed since the first moment, but at the instant of its creation the universe was expanding infinitely rapidly. No force caused it to explode in this way, it simply started with an initial expansion. Had the explosive vigour been less extreme, then gravity would soon have overwhelmed the dispersing material, reversing the expansion and engulfing the entire cosmos in a catastrophic implosion, producing something rather like a black hole. But the bang was big enough to enable the universe either to escape its own gravity and go on expanding for ever under the impetus of the initial explosion, or at least to survive for many thousands of millions of years before succumbing to implosion and annihilation.

The trouble with this traditional picture is that it is no explanation for the big bang. Once again, a fundamental feature of the universe is merely attributed to an *ad hoc* initial condition. Why did the universe not explode more violently still, in which case it would be expanding much faster today? Alternatively, why is it not expanding much slower, or even contracting by now?

Careful measurement puts the rate of expansion very close to a critical value at which the universe will just escape its own gravity and expand for ever. A little slower, and the cosmos would collapse, a little faster and the cosmic material would have long ago completely dispersed. How delicately the rate of expansion has been 'fine-tuned' to fall on this narrow dividing line between two catastrophes. If at time 1 s (by which time the pattern of expansion was already firmly established) the expansion rate had differed from its actual value by more than 10^{-18}, it would have been sufficient to throw the delicate balance out. The explosive vigour of the universe is thus matched with almost unbelievable accuracy to its gravitating power. The big bang was evidently, an explosion of exquisitely arranged magnitude.

The rate of expansion is only one of several apparent cosmic 'miracles'. Another concerns the pattern of expansion. The present universe is extraordinarily uniform on the large scale, in the way that matter and energy are distributed. The primeval heat radiation which bathes the universe arrives at Earth with a uniform temperature in every direction accurate to one part in ten thousand.

The large-scale uniformity of the universe continues to be preserved with time as the universe expands. It follows that the expansion itself must be uniform to a very high degree. So not only did the universe commence with a bang of a quite precise magnitude, it was a highly orchestrated explosion, a simultaneous outburst of exactly uniform vigour everywhere and in evey direction.

The extreme improbability that such a coherent, synchronized eruption would occur spontaneously is exacerbated by the fact that, in the traditional big bang theory, the different regions of the primeval cosmos would have been causally isolated. On account of the theory of relativity, no physical influence can propagate faster than light. Consequently, different regions of the universe can come into causal contact only after a period of time has elapsed. For example, at 1 s after the initial explosion, light can have travelled at most one light-second which is 300000 km. Regions of the universe separated by greater than this distance could not, at 1 s, have exercised any influence on each other. But at that time, the universe we observe today occupied a region of space at least 10^{14} km across. It must therefore have been made up of some 10^{27} causally separate regions, all of them, nevertheless, expanding at exactly the same rate. Even today, when we observe the cosmic heat radiation coming from opposite sides of the sky, we are receiving identical thumbprints from regions of the universe that are separated from each other by 90 times the distance that light could have travelled at the time the heat radiation was emitted towards us.

The large-scale uniformity of the universe is all the more mysterious because, on a smaller scale, the universe is *not* uniform. The existence of galaxies and galactic clusters indicates a departure from exact uniformity, a departure which is, moreover, of the same magnitude and scale everywhere. Because gravity tends to amplify any initial clumping of material, the degree of non-uniformity required to produce galaxies was far less during the big bang than it is today. In spite of this, some degree of irregularity must have been present in the primeval phase or galaxies would never have started to form. In the old big bang theory these early irregularities were also explained away as initial conditions. Thus, we were required to believe that the universe began in a peculiar state of extraordinary but not quite perfect order.

This genesis paradox can be summarized as follows: with gravitational attraction the only cosmic force available, the big bang must simply be accepted as god-given, an event without a cause, an assumed initial condition. Furthermore, it was an event of quite astonishing fidelity, for the present highly structured cosmos could not have arisen unless the universe was set up in just the right way at the outset.

The search for antigravity

Although a resolution of the genesis paradox has only been achieved in the past few years, traces of the essential idea go back to a time before the expansion of the universe or the big bang theory were known. Even Newton realized that there was a deep puzzle about the stability of the cosmos. How can the stars just hang out there in space unsupported? The universal force of gravity, being attractive, ought to cause the entire collection of stars to plunge in on itself.

To escape from this absurdity, Newton argued that if the universe collapsed under its own gravity every star would be obliged to fall towards the centre of the stellar assemblage. But suppose the universe were infinite, with stars distributed on average uniformly throughout infinite space. There would then be no overall centre towards which the stars could fall; in an infinite universe every region is identical to every other. Any given star would receive gravitational pulls from all its neighbours, but these would average out in their different directions, and so there would be no systematic net force to convey a star towards any particular place of general congregation.

Einstein replaced Newton's gravitation theory 200 years later. His first paper on cosmology, published before Hubble's discovery of the expanding universe, presumed, like Newton, that the cosmos was static. His solution of the stability problem was, however, much more direct. Einstein believed that to prevent the universe from imploding under its own gravity there had to be another cosmic force to counteract the gravitational force. This new force would have to be repulsive rather than attractive. In this respect it might be regarded as 'antigravity', although 'cosmic repulsion force' is a more accurate description.

Although the idea of a repulsive force pushing against the gravity of the universe is simple enough, the actual properties of the force are peculiar. We do not notice any such force on the Earth, nor has any hint of one ever been found. If a cosmic repulsion force exists, it must not act conspicuously at close range but accumulate in strength over astronomical distances. Behaviour of this sort was contrary to all previous experience of forces, which tend to be strong nearby and to weaken with distance. Electric and gravitational forces, for example, fall steadily towards zero in accordance with the inverse square law.

The cosmic repulsion found by Einstein should not be thought of as a fifth force of nature. Rather, it is a weird offshoot of gravity itself. In fact, the effects of cosmic repulsion can be attributed to ordinary gravity if the source of the gravitational field is chosen to be a medium with rather unusual properties. A familiar material medium, such as a gas, will exert a pressure, but the hypothetical cosmic medium being discussed here is required to possess a *negative* pressure, or tension.

We can envisage the repulsion, therefore, either as a sort of adjunct of gravity, or as caused by the ordinary gravity of an invisible fluid medium with negative pressure filling all of space. The repulsive effect is due to the *gravity* of the medium, not its mechanical action. In any case, mechanical forces arise from pressure differences, not pressure as such and the medium is supposed to fill all of space. It could not be confined to a vessel. In fact, an observer immersed in the medium would not perceive any tangible substance at all. However, Einstein believed that he had a convincing model of a universe held in equilibrium between the attractive force of gravity and the newly discovered cosmic repulsive force. He calculated the strength needed for the repulsive force to balance the gravity of the universe and confirmed that the repulsion would be so slight within the solar system, and even the galaxy, that we would never have spotted it observationally. For a while it seemed that an age-old puzzle had been brilliantly solved.

Then things began to go wrong. First there was a problem about stability. The essential idea was to match the forces of attraction and repulsion precisely. But like many balancing acts this one turns out to be a delicate affair. If, for example, Einstein's static universe were to expand a fraction, the attractive force of gravity (which diminishes with separation) would go down a bit, while the cosmic repulsion force (which increases with distance) would go up. This would lead to an overbalance, with the repulsion winning out and forcing a still greater expansion, and leading to the eventual runaway distension of the universe under an all-dominating repulsion. On the other hand, if the universe shrank a little, the gravitational force would go up and the repulsion would go down, causing gravity to win out, and the universe would then shrink faster and faster towards the total collapse that Einstein had sought to avoid. Thus, the slightest hiccup, and the carefully balanced equilibrium would fail, spelling cosmic disaster.

Then, in 1927, Hubble discovered the expansion of the universe. All balancing acts were thereby rendered obsolete. it was immediately apparent that the universe avoids *implosion* because it is engaged in *explosion*. Had Einstein not been sidetracked with the repulsive force he would surely have made this deduction theoretically and thus have predicted the expansion of the universe a decade before its discovery by astronomers. In the event, Einstein abandoned the cosmic repulsion force in disgust. But that was not the end of the story.

Cosmic repulsion was invented by Einstein to solve a non-existent problem, namely how to explain a static universe. But although astronomical observations do not reveal cosmic repulsion at work, they cannot prove it is non-existent. It may simply be too weak to have shown up yet.

Einstein's field equations, although admitting a repulsive force in a natural way, make no restriction on the *strength* of the force. Einstein was free to postulate that the strength was precisely zero, thus eliminating the repulsion altogether. But there was no compelling reason to do this. Other scientists were happy to retain the repulsion, in the absence of evidence to the contrary, they reasoned, nobody is justified in setting the force to zero.

The consequences of retaining the repulsion force in the expanding universe scenario are easily worked out. Early in the life of the cosmos, when the universe is compressed, the repulsion can be ignored. During this phase, the gravitational attraction slows the pace of expansion, in the same way that a missile fired vertically is slowed by the Earth's gravity. If it is assumed that the universe starts out expanding rapidly, then gravity acts to steadily reduce the rate to the value observed today. With time, the gravitational force weakens as the cosmic material disperses. By contrast, the cosmic repulsion grows because the galaxies move farther apart. Eventually, the repulsion force comes to exceed the gravitational attraction, and the expansion rate begins to get faster and faster. Thereafter, the universe is dominated by the cosmic repulsion and spends all of eternity in runaway expansion.

Astronomers have reasoned that this unusual behaviour, in which the universe first slows up and then accelerates again, ought to be evident in the observed motion of the galaxies. Careful astronomical observations have failed to provide any convincing evidence for such a turn-about, although claims have been made to the contrary.

The idea of a universe caught by runaway expansion had been mooted by the Dutch astronomer Wilhelm de Sitter in 1916, who argued that if the universe were devoid of ordinary matter, then the usual attractive force of gravity would be absent, and the cosmos would come under the sole influence of the repusion. This would make the universe expand. This idea of expanding empty space was considered little more than a curiosity at the time, although it turned out to be remarkably prophetic.

The fact that astronomers do not see a cosmic repulsive force at work does not logically imply that the force is non-existent. It may be too weak to be detected with present instrumentation. All observations contain a level of error, and only an upper limit on the strength of the force can be obtained. Set against this, it could be argued that the laws of nature would be simpler if cosmic repulsion were absent. This inconclusive debate on the existence of 'antigravity' had been grinding on for many years when suddently a new twist occurred that gave the subject an unexpected immediacy.

Inflation: the big bang explained

The conclusion that if a cosmic repulsive force exists, it must be very weak and far too weak to have had any significant effect on the big bang assumes that the strength of the repulsive force does not change with time. In Einstein's day no one considered the possibility that could change as the universe expands. Had such a possibility been entertained, then one could then conceive of a scenario in which, under the extreme conditions of the early universe, cosmic repulsion momentarily dominated gravity causing the universe to explode, before fading into insignificance.

This general scenario is precisely what has come out of recent work on the behaviour of matter and forces in the very early universe. It is now clear that a huge cosmic repulsion is an inevitable by-product of the activities of the superforce. The 'antigravity' that Einstein threw out of the door has come back in through the window.

The key to understanding the re-discovered cosmic repulsion is the nature of the quantum vacuum. We have seen how such a repulsion can be produced by a bizarre invisible medium which looks identical to empty space but which possesses a negative pressure. Physicists now believe that that is exactly how a quantum vacuum would be.

When physicists began to study the quantum theory of fields, they discovered that a vacuum was not just empty space devoid of substance and activity. Quantum physics seemed capable of playing tricks even in the absence of any quantum particles. The law of energy conservation can be suspended by quantum effects for a very short interval of time. During this brief duration energy can be 'borrowed' for all manner of purposes, one of which is to create particles. Any particles so produced will be short-lived, because the energy tied up in them has to be repaid after a minute fraction of a second. Nevertheless, particles can pop out of nowhere, enjoying a fleeting existence, before fading once again into oblivion. This evanescent activity cannot be prevented. Though space can be made as empty as it can possibly be, there will always be a host of these temporary particles. The temporary 'ghost' particles cannot be seen, even though they may leave physical traces of their brief existence. They are a form of 'virtual' particle, similar to messenger particles, but with nothing on the 'ends of the line' to send or receive the message. They travel from emptiness to emptiness, witnesses to the existence of a force field, but with nothing permanent to push against.

What might appear to be empty space is, therefore, a seething ferment of virtual particles. A vacuum is not inert and featureless, but alive with throbbing energy and vitality. A 'real' particle such as an electron must always be viewed against this background of frenetic activity. When an electron moves through space, it is actually swimming in a sea of ghost particles of all varieties—virtual leptons, quarks, and messengers. The presence of the electron will distort this irreducible vacuum activity, and the distortion in turn reacts back on the electron. Even at rest, an electron is not at rest: it is being continually assaulted by all manner of other particles from the vacuum. It is important to appreciate that, at the quantum level of description, the vacuum is the dominant structure. What we call particles are only minor disturbances bubbling up over this background sea of activity.

In the late 1970s, the unification of the four forces was followed by a drastic re-appraisal of the physical nature of the vacuum. The theory suggested that all this vacuum energy could arrange itself in more than one way. The vacuum could become excited and adopt a number of states of very different energy, in the same way that an atom can be excited to higher energy levels. These several vacuum states would look identical if we could view them, but they possess very different properties.

First of all, the energy involved leaps by huge amounts from one vacuum state to another. In the grand unified theories, to take an example, the gap between the least and greatest vacuum energy is almost incomprehensibly large. Alongside of which are equally enormous changes in the pressure of the vacuum states. But the pressures are all *negative*. The quantum vacuum behaves exactly like the previously hypothetical medium which produces cosmic repulsion, only this time the numbers are so big that the strength of the repulsive force is 10^{120} times greater than Einstein needed to prop up a static universe.

Suppose that, in the beginning the universe found itself in an excited vacuum state. In this state, the universe would be subject to a cosmic repulsion force of such magnitude that it would cause headlong expansion at a huge rate. In fact, the universe would resemble de Sitter's model. The difference is that, whereas de Sitter envisaged a universe sedately expanding over an astronomical time-scale, the de Sitter phase driven by the false quantum vacuum is far from sedate. A typical region of space would double in size every 10^{-34} s or so!

The way in which this hyper-expansion proceeds is distinctive: distances increase in size exponentially fast. This means that every 10^{-34} s every region of the universe doubles its size, and then goes on doubling again and again in a progression. This type of runaway expansion has been dubbed 'inflation' by Alan Guth, who invented the idea in 1980. Under the impact of the exceedingly rapid and accelerating expansion, the universe would have soon have found itself swelling explosively fast. This was the big bang. Somehow, the inflationary phase has to terminate. As with all excited quantum systems, the false vacuum is unstable and will tend to decay. When that happens, the repulsion force disappears. This would have put a stop to inflation, bringing the universe under the control of ordinary, attractive gravity. The universe would have continued to expand, of course, from the initial impetus imparted by the inflationary episode, but at a steady falling rate. The only trace that now remains of the cosmic repulsion is this dwindling expansion.

According to the inflationary scenario, the universe started out in a vacuum state, devoid of matter or radiation. Even if matter and radiation were present initially, all traces would soon have been eradicated because the universe swelled by such an enormous fact during the inflationary phase. During this incredibly brief phase, the region of space which today forms the entire observable universe grew from one-thousand-millionth of the size of a proton to several centimetres. The density of any pre-existing material would have fallen essentially to zero.

At the end of inflation, then, the universe was empty and cold. As soon as inflation ceased, however, the universe was suddenly filled with intense heat. This flash of heat which illuminated the cosmos owed its origin to the huge reserves of energy locked up in the false vacuum. When the false vacuum decayed, its energy was dumped in the form of radiation, which instantly heated the universe to about 10^{27} K, hot enough for GUT processes to occur. From this point on the universe evolved according to the standard hot big bang theory. The heat energy created matter and antimatter, the universe began to cool, and in a succession of steps all the structure we observe today began to 'freeze' out.

The problem of what caused the big bang is therefore solved by the inflationary theory: empty space itself exploded under the repulsive power of the quantum vacuum. But the colossal energy of the primeval explosion that went to generate all the matter and radiation we now see in the universe had to come from somewhere. We will not have explained the existence of the universe until we have traced the source of the primeval energy.

One of the fundamental laws of physics is the law of *conservation* of energy, which says that although you can change energy from one form to another, the total quantity of energy stays fixed.

If energy cannot be created or destroyed, how did the primeval energy come to exist? Was it simply injected at the beginning of time, another *ad hoc* initial condition? If so, why does the universe contain the amount of energy that it does?

The inflation theory is one possible scientific answer to this mystery. According to the theory, the universe started out with essentially zero energy, and succeeded in conjuring up the lot during the first 10^{-32} s. The key to this miracle is that the law of conservation of energy *fails* when applied to the expanding universe.

In fact, we have already encountered this point. The cosmological expansion causes the temperature of the universe to fall. The radiant heat energy that was so intense in the primeval phase had dwindled to a temperature close to absolute zero. Where has all that heat energy gone? The answer is, that in a sense it has depleted itself by helping the universe to expand, adding its pressure to the explosive violence of the big bang. If you expand an ordinary gas, its internal energy must fall to pay for the work done. In stark contrast to this conventional behaviour, the cosmic repulsion behaves like a fluid with *negative* pressure. When a negative-pressure fluid is expanded, its energy goes *up* rather than down. This is what happened in the inflationary period, when the cosmic repulsion drove the universe into accelerated expansion. The total energy of the vacuum kept on rising until, at the cessation of the inflationary era, it had accumulated to a huge amount. As soon as inflation stopped, this energy was released in a single great burst, generating all the heat and matter that eventually emerged from the big bang. From then on, the conventional positive-pressure expansion took over, and the energy began to decline again.

The creation of the primeval energy has an air of magic to it. The vacuum, with its weird negative pressure, on the one hand produces a powerful repulsive force, bringing about its own accelerating expansion; on the other hand, that very expansion goes on boosting the energy of the vacuum. The vacuum essentially pays itself vast quantities of energy. It has an inbuilt instability to continue expanding and generating unlimited quantities of energy for free. Only the quantum decay of the false vacuum stops the bonanza.

The vacuum is nature's miraculous jar of energy. There is, in principle, no limit to how much energy can be self-generated by inflationary expansion. It is a revolutionary result at total variance with the centuries-old tradition that 'nothing can come out of nothing'. Christians have long believed that God created the universe out of nothing, but the possibility that all the cosmic matter and energy might appear spontaneously as a result of purely physical processes would have been regarded as utterly untenable by scientists only a decade ago.

There is an alternative way of looking at the creation of energy by the expanding universe. Because gravitational forces are normally attractive, it is necessary to do work to pull matter apart against its own gravity. This means that the gravitational energy of a collection of bodies is negative; if more bodies are added to the system, energy is released and the gravitational energy becomes more negative to pay for it. In the context of the inflationary universe, the appearance of heat and matter could be viewed as exactly compensated by the negative gravitational energy of the newly created mass, in which case the total energy of the universe is zero, and no net energy has appeared after all! Attractive though this way of looking at the creation may be, it should not be taken too seriously because the whole concept of energy has dubious status as far as gravity is concerned.

Successes of inflation

Once the basic idea had been mooted that the universe underwent an early period of extremely rapid expansion, it became apparent that the scenario provides an elegant explanation for many of the previously *ad hoc* features of big bang cosmology.

I have mentioned the way in which the strength of the explosion exactly matched the gravitational power of the cosmos such that the expansion rate today lies very close to the borderline between re-collapse and rapid dispersal. A crucial test of the inflationary scenario is whether it produces a big bang of this precisely matched magnitude. Because of the nature of exponential expansion—the characteristic feature of the inflationary phase—the explosive power is automatically adjusted to yield exactly the right value corresponding to the universe just escaping its own gravity. Inflation can give no other expansion rate than the one that is observed.

The large-scale uniformity of the universe is also immediately explained by inflation. Any irregularities initially present in the universe would have been stretched to death by the enormous distension. With regions of space being expanded by factors of 10^{50}, any prior disorder would be diluted to insignificance.

Complete uniformity would be incorrect, because a small degree of clumping was necessary in the early universe to account for the present existence of galaxies and galactic clusters. Astronomers hoped that the existence of galaxies might be explained as a result of gravitational aggregation since the big bang. A cloud of gas will tend to contract under its own gravity and then fragment into smaller clouds, which in turn fragment into still smaller clouds, and so on. It is possible to imagine the gas emerging from the big bang uniformly distributed, but purely by chance accumulations becoming overdense here and there and underdense elsewhere. Gravity would reinforce this tendency, causing the enhanced regions to grow stronger and suck in more material, and then to shrink and successively fragment, with the smallest fragments becoming stars. One would then end up with a hierarchy of structure, with stars clustered into groups, which in turn cluster into galaxies and galactic clusters.

Unfortunately, the growth of galaxies by this mechanism would take much longer than the age of the universe if there were no irregularities present in the gas at the outset, because the shrinking and fragmenting process is in competition with the expansion of the universe, which tries to disperse the gas. In the old version of the big bang theory, the seeds of galaxies were already built into the structure of the universe when it was created. Moreover, these initial irregularities had to be of just the right magnitude: too small and galaxies would never form, too large and the overdense regions would collapse into huge black holes instead. We are at a loss to know why the galaxies are the sizes they are, or why the clusters contain the numbers of galaxies that they do.

A better explanation for galactic structure is based on the inflationary scenario. Inflation occurs while the quantum state of the universe is hanging in the unstable 'false' vacuum state. Eventually the false vacuum decays, its excess energy going into heat and matter. At this point, the cosmic repulsion disappears and inflation ceases. However, the decay of the false vacuum does not occur at exactly the same instant throughout space. Some regions of the universe will decay slightly faster than others. In these regions, inflation will end sooner and irregularities can hopefully act as seeds or centres for gravitational clumping that eventually lead to galaxies and galactic clusters.

In Guth's original version of the inflationary scenario, the false vacuum decayed abruptly into the 'true' vacuum, the lowest energy vacuum state, which we identify with empty space. The way in which this change occurred was regarded as similar to a phase transition such as from a gas to liquid. Bubbles of true vacuum were envisaged as forming at random in the false vacuum, and then expanding at the speed of light to encompass greater and greater volumes of space. To enable the false vacuum to live long enough for inflation to work its magic, the two states were separated by an energy barrier through which the system was obliged to 'quantum tunnel'. This model suffered from a major shortcoming, however: all the energy released from the false vacuum was found to be concentrated in the bubble walls, and there was no mechanism to distribute it evenly through the interior of bubbles. When the bubbles collided and coalesced, the energy would end up in tangled sheets. The resulting universe would contain severe irregularities, and the work of inflation in achieving large-scale uniformity would be ruined.

In the new theory, there is no tunnelling between the two vacuum states, instead parameters are chosen so that the decay of the false vacuum is very slow, giving the universe time to inflate. When decay eventually occurs the energy of the false vacuum is released throughout the 'bubble', which quickly heats up to 10^{27} K. It is assumed that the entire observable universe is contained within a single bubble. Thus, on an ultra-large scale the universe may be very irregular, but our own region (and much more beyond) lies within a domain of quiescent uniformity.

Guth's original reason for inventing the inflationary scenario was to address the absence of magnetic monopoles. The standard big bang theory predicts that a superabundance of monopoles would have been created in the primeval phase. It also happens that these monopoles are likely to be accompanied by other bizarre objects known as 'strings' and 'sheets' which are their one- and two-dimensional analogues. The problem was how to rid the universe of these undesirable entities. Inflation solves the monopole and related problems automatically because the enormous swelling of space effectively dilutes them to zero density.

The self-creating universe

How did the universe arrive in the false vacuum state in the first place? What happened *before* inflation?

A completely satisfactory scientific account of the creation would have to explain how space (strictly spacetime) came to exist, so that it might then undergo inflation. Some scientists assume either that space always existed, or that its creation lies beyond the scope of science. A few, however, believe that it is possible to discuss how space in general, and the false vacuum in particular, might have come out of literally nothing as a result of physical processes that are, in principle, amenable to study.

The belief that nothing can come out of nothing has only recently been challenged. It is certainly true that in the familiar world of experience objects usually owe their existence to other objects. The Earth was formed from the solar nebula, the solar nebula from the galactic gases, and so on. Nevertheless, can we conceive of physical objects, or even the entire universe, coming into existence out of nothing?

In such conjectures the quantum factor provides the key. The central feature of quantum physics, is the disintegration of the cause–effect link. In classical physics, the science of mechanics exemplified the rigid control of causality. The activity of every particle, was considered to be legislated in detail by the laws of motion. A body was understood to move continuously in a well-defined way according to the pattern of forces acting upon it. The laws of motion embodied the link between cause and effect in their very definition, so that the entire universe was supposed to be regulated in every minute respect by the existing pattern of activity. The universe, according to this view, is forever unfolding along a pre-ordained pathway.

Quantum physics wrecked this orderly, yet sterile scheme. Physicists learned that at the atomic level, matter and motion are vague and unpredictable. Particles can behave erratically, rebelling against rigidly prescribed motions, turning up in unexpected places without discernible reason and even appearing or disappearing without warning.

Causality is not completely absent in the quantum realm, but it is faltering and ambiguous. If an atom, for example, is excited by a collision with another atom, it will usually return quickly to its lowest energy state by emitting a photon. The coming-into-being of the photon is, naturally, a consequence of the atom's being excited in the first place. We can certainly say that the excitation caused the creation of the photon. In that sense cause and effect remain linked. Nevertheless, the actual moment of creation of the photon is unpredictable; the atom might decay at any instant. Physicists can compute the expected, or average, delay before the photon appears, but they can never know in any individual case when this event will happen. Perhaps it is better to say that the excitation of the atom 'prompts' rather than causes the photon to come into being.

Why do we find the idea of an object abruptly appearing from nothing so incredible? Perhaps the answer lies with familiarity. *If we could* actually observe the behaviour of atoms directly with our sense organs, rather than through the intermediary of special instruments, we should frequently see objects appearing and disappearing without well-defined reasons.

The closest known instance to the idea of creation out of nothing occurs if an electric field can be made strong enough. At a critical field strength, electrons and positrons start appearing out of nowhere in an entirely random way. This spontaneous creation can be considered as a bizarre type of radioactivity, in which it is empty space—the vacuum—which decays.

Although the decay of space is difficult to achieve using an electric field, an analogous process involving gravity might well occur naturally. Near the surface of black holes, gravity is so intense that the vacuum sizzles with a continual stream of newly created particles. This is the black hole radiation discovered by Stephen Hawking. Gravity is ultimately responsible for creating the radiation, but it does not cause it: no given particle has to appear at any particular place and time as a result of gravitational forces.

Gravity is only a warping of spacetime, so we could say that it is spacetime that induces the creation of matter.

The spontaneous appearance of matter out of empty space comes close to the spirit of the creation *ex nihilo*. For the physicist, however, empty space is a far cry from nothing: it is very much part of the physical universe. To answer the ultimate question of how the universe came into existence it is not sufficient to assume that empty space was there at the outset. We have to explain where space itself came from. In a sense *space* is being created around us all the time. The expansion of the universe is nothing but a continual swelling of space. Space is rather like super-elastic in that it can go on stretching forever (as far as we know) without 'snapping'.

The stretching and warping of space also resembles elastic inasmuch as the 'motion' of space is subject to laws of mechanics in the same way as matter. These are the laws of gravity. Just as the quantum theory applies to the activities of matter, so it applies to space and time. If quantum theory allows particles of matter to pop into existence out of nowhere, could it also, when applied to gravity, allow space to come into existence out of nothing? And if so, should the spontaneous appearance of the universe 18000 million years ago occasion such surprise after all?

The concept of quantum cosmology, superficially, seems to be a contradiction in terms. Quantum physics deals with the smallest systems, while cosmology is the study of the largest. Nevertheless, the universe was once very shrunken, and there must have been a time when quantum effects were important. Calculations suggest that quantum physics cannot be ignored at the GUT era (10^{-32} s) and would probably have dominated everything at the Planck era (10^{-43} s). It was at some moment between these two epochs when, according to theorists such as Vilenkin, the quantum universe erupted into existence. In the words of Sidney Coleman, "We make a quantum leap from nothing into time." Spacetime, it seems, is a fossil from this era.

Coleman's "quantum leap" could be described as a form of 'tunnelling'. In the original inflation theory, the false vacuum state was required to tunnel into the true vacuum state through an energy barrier. In the case of the spontaneous appearance of the quantum universe out of nothing, however, our intuition is strained to the limit. One end of the 'tunnel' represents the physical universe of space and time, which 'arrived' by quantum tunnelling from nothing, and so the other end of the tunnel must be 'nothing'! Perhaps it would be better to say that there is only one end to the tunnel; the other end does not exist.

How can we account for the fact that the cosmos was created in a false vacuum state? Had the newly created spacetime been in the true vacuum, inflation would never have occurred, the big bang would have been reduced to a whimper, and spacetime would have shrunk back out of existence after a fleeting instant, devoured by whatever quantum activity produced it initially. Without being in the false vacuum the universe could never have made concrete its ephemeral existence. A false vacuum state may have been favoured by the extreme conditions prevailing at the time. For example, if the universe was created at a high enough initial temperature and then cooled, it might have been stranded in a false vacuum.

Whatever the truth, quantum physics offers the only branch of science in which the concept of an event without a cause makes sense. When the subject at issue concerns spacetime it is meaningless to talk about a cause in the usual sense. Causation is rooted in the notion of time, and so any ideas about an agency creating time, or causing time to come into existence, must appeal to a wider concept of causality than is familiar in science.

Dr. Basil J. Hiley

Basil Hiley is Reader in Theoretical Physics at Birkbeck College, University of London.

After graduating in Physics, and PhD in Theoretical Physics in 1962, he became Lecturer in Physics at Birkbeck College, until his present appointment in 1980.

His research activities are concerned with solid state physics, with special interests in co-operative phenomena, liquid state physics and polymer physics, the conceptual basis of quantum mechanics (with special emphasis on the quantum potential model and non-locality), the nature of space-time and quantum aspects of gravity (with specific reference to Clifford algebras, twistors, de Rahm cohomology, and supersymmetries).

He is author of over fifty technical research papers in specialist journals, and a number of non-technical reviews of a more general nature.

The elusive fields of non-locality

Anyone who has looked at quantum phenomena and studied the mechanics by which they are described, cannot but be surprised and puzzled. The phenomena themselves seem to defy our intuition derived from classical mechanics. You cannot use billiard-ball analogies. If you look at the mathematical formalism you see 'wave functions', 'operators', 'observables' and so on, while the classical notions of position, momentum, angular momentum, *etc.*, appear in a rather strange way. The whole thing seriously challenges our outlook on reality and what I would like to do is to discuss the subject in a somewhat unorthodox way, which I hope will reveal some possible insights into the nature of reality.

Our traditional view of nature was originated, perhaps, by Democritus, described vividly by Lucretius, and finally confirmed by the empiricists of the eighteenth/nineteenth centuries (such as Dalton and Avogadro). Put simply, it states: *Complexity is to be understood through objects-in-interaction.* These objects have been first atoms, then electrons, protons, neutrons and now they are quarks or even preons. Physicists continually search for the ultimons and their laws of interaction hoping these will be simple, in space-time and, above all, local. Hence, *simplicity will give rise to complexity.*

But then quantum mechanics came along and the hoped-for simplicity seemed to become more and more illusive. We do not have to go as far as quarks or preons. Even atoms themselves behave strangely, although it is easier experimentally to produce quantum effects with electrons and neutrons. For example, we all know of Young's slits and the interference pattern produced by light. If you pass electrons or neutrons through slits of the appropriate size, these build up into an 'interference' pattern of individual arrivals. Again, we know that these particles pass through barriers that, classically, they should not be able to penetrate. It is like leaving your car securely locked in the garage at night and finding it down the road next morning! How can such things happen? What new concepts are needed to understand this sort of behaviour? Niels Bohr gave a surprising answer—*none*! The mathematics takes care of everything—but explains nothing. As Nobel prize winner Murray Gell-Mann[1] puts it,

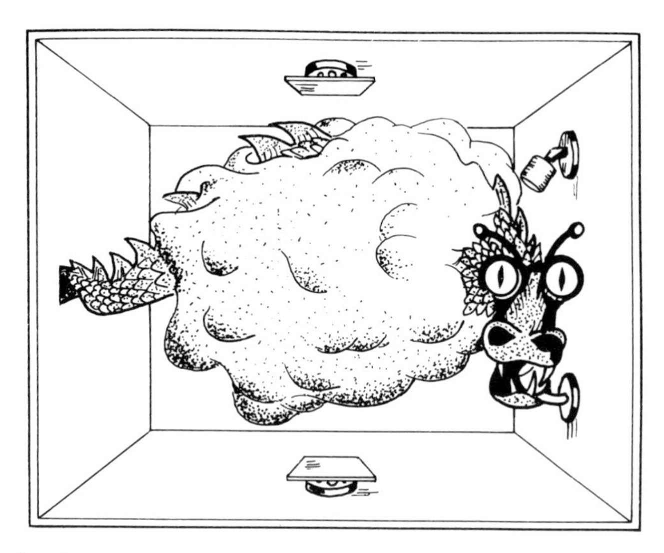

Figure 1 Smoky dragon as symbol of Bohr's elementary quantum phenomenon. (Drawing by Field Gilbert, Austin 1983).

"Quantum mechanics, that mysterious, confusing discipline, which none of us really understands but which we know how to use".

The mathematics uses wavefunctions $\psi(\mathbf{r},t)$ which are sometimes regarded as 'state-functions' describing a particle in a given situation. In the usual approach to quantum mechanics, these functions develop in time following Schrödinger's equation but it is not possible to describe how the particle itself moves. There is an essentially ambiguous relation between the particle and the wave function. Wheeler[2] has a very vivid way of illustrating this point in terms of a smokey dragon. The tail of the dragon is sharp where the particle starts and its mouth where it bites the counter is sharp but in between . . . who knows? (see Figure 1.)

The difficulty is that there is no notion of an *independent actuality* in the usual approach. That is, there is no actual process by which the particle at one point passes over into another. All that seems to be possible is to say what happens as a result of a measurement, but between measurements one can say nothing. The theory is, therefore, about *appearances or phenomena*. As Bohr[3] puts it "As a more appropriate way of expression, I advocate the application of the word phenomenon exclusively to refer to the observations obtained under specified circumstances, including an account of the *whole* experimental arrangement." Bohr's[4] reason for this is that "in quantum mechanics we are not dealing with an arbitrary renunciation of a more detailed analysis of atomic phenomena, but with a recognition that such an analysis is in principle excluded."

But why should Bohr exclude analysis in principle? He argued that the uncertainty principle implies that you cannot say precisely what belongs to the observed and what belongs to the observing instruments. He uses the analogy of a blind man trying to obtain a mental image of a room. As we know, with a rigid stick a blind person can arrive at a remarkably accurate 'picture' of the room, but give him a flexible rubber stick and he is in trouble. There is no longer a sharp separation between the observer and the observed. The situation is the same in quantum theory, according to Bohr. Analysis is out—wholeness is in. Thus, Bohr totally rejected the traditional view of trying to understand qunatum phenomena and replaced it with the notion of unanalysable wholeness.

I do not think many physicists really believe Bohr's position. They certainly do not act as if they believe it! Surely we can, in some way, reformulate quantum mechanics that will enable us to talk about an actual process? The inevitable answer will be "but we know you cannot do that, von Neumann and others have shown it to be impossible". But they are wrong. You can reformulate quantum mechanics and account for all the quantum phenomena correctly. The idea is very simple. One assumes that particles exist and that they have precise positions and momenta, even though these cannot be measured simultaneously. One further assumes that they also have a field-like quality which is described by the wave function $\psi(\mathbf{r},t)$ and satisfies Schrödinger's equation. We can quickly derive the defining equations. Put $\psi = Re^{iS/\hbar}$ in Schrödinger's equation and separate it into real and imaginary parts. It is easy to show the following:[5]

$$\frac{\partial S}{\partial t} + \frac{(\nabla S)^2}{2m} + V + Q = 0 \tag{1}$$

where $$Q = \frac{-\hbar^2}{2m}\frac{\nabla^2 R}{R}$$

and $$\frac{\partial R}{\partial t} + \text{div}\left[R^2\frac{\nabla S}{m}\right] = 0 \tag{2}$$

Equation (1) can also be written in a more familiar way

$$\frac{d\mathbf{p}}{dt} = -\nabla(V + Q) \tag{3}$$

where $\mathbf{p}$ is the momentum of the particle. This is just Newton's law of motion for a classical particle, except for an additional potential, Q. Thus when Q is negligible, the particle behaves like a classical particle. The extra potential Q changes its behaviour in such a way that enables us to obtain the quantum results for an ensemble of individual particles. For this reason, we call it the quantum potential.

For a given wave function, we can calculate Q and solve equation (3) to find a trajectory for the particle. Thus, in this approach there is an actuality, namely, a particle following a well defined trajectory. If we assume that $R^2(\mathbf{r},t)$ is the probability that the particle will be at a certain point in space-time, then equation (2) ensures a conservation of probability. This means that if we take an ensemble of initial positions and momenta commensurate with the probability distribution derived from the initial wave function then the final probability distribution will be identical to the probability distribution calculated from the final wave function. This ensures that we get an exact agreement with quantum mechanics.

We can now account for quantum phenomena in a mathematical language that is much closer to that used in classical physics. This does not mean that we have returned to classical physics, but we will be able to see more clearly the differences between classical and quantum physics.

The best way to get an insight into how the quantum potential works is to look at a specific example. I will take the two-slit experiment. Here we must resort to numerical calculations which were carried out by two of our former research students, C. Philippidis and C. Dewdney.[6] In Figure 2, an ensemble of trajectories is shown. The two slits are on the left, while the

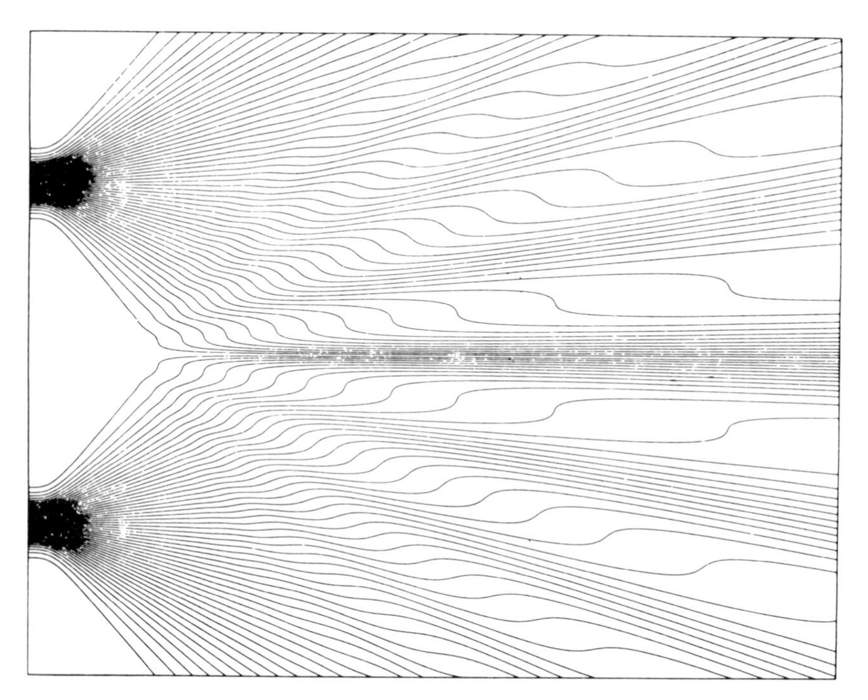

Figure 2 Particle trajectories in two-slit experiment for electrons. The slits are on the left-hand side, the fringes are formed on the right.

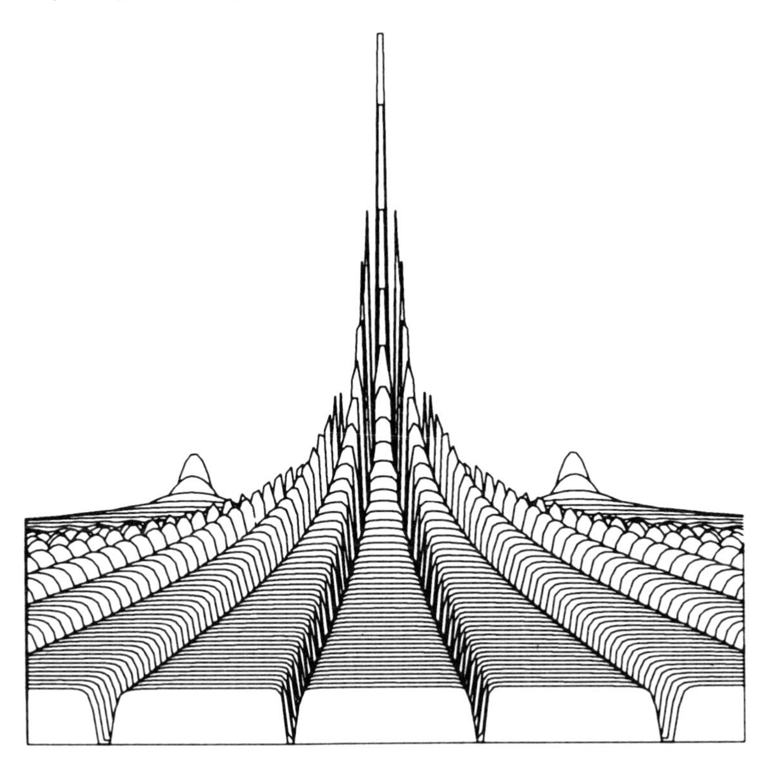

Figure 3 The quantum potential for the two-slit experiment. The slits are at the back while the fringes are formed in the foreground of the figure.

screen is on the right of the figure. Each particle will travel along one of the trajectories depending on its initial position at the slits. One immediately notices that after the particles have passed through the slit, they no longer travel in a straight line as we would expect if classical mechanics were operating. Instead, they show a series of kinks and it is these kinks that give rise to the bunching of the particles. Thus we see how the 'fringes' arise out of an ensemble of individual events.

But why should these kinks appear? Clearly, they must arise from the quantum potential and to see this in more detail we must calculate this potential. The result is shown in Figure 3. Note that the potential contains a number of plateaus and valleys. As the force is calculated from the derivative of the potential, the particles only experience a significant force as they cross the valleys. These valleys coincide exactly with the kinks in the individual trajectories. The quantum potential, therefore, organizes the behaviour of the individual particles in such a way as to produce the characteristic quantum effects.

Let us examine the quantum potential in more detail as it is solely responsible for the quantum behaviour. The first thing to notice is that its shape is unlike any known classical potential. It does not have a point source and it is not radiated. It contains information about the slit widths, their separation, their shape, the particles' momentum, and so on. It is also still active a long way from the slits. One of the reasons it can act over large distances is that the potential depends on the *form* of the wave, not its amplitude. Thus even if the amplitude is very small, the quantum potential can produce a large effect. This is very different from the classical situation where the effects are proportional to the amplitude and, therefore, the force falls off with distance.

It is often suggested that we should be able to verify experimentally the appearance of kinks in the trajectories. However, this is not possible, because whenever we introduce an apparatus to measure the effect, a new quantum potential arises and this new potential destroys the old potential, so that the particles no longer follow the original trajectories. Thus, even in the quantum potential approach we support a point that Bohr stressed, namely, that the phenomenon depends on the whole experimental arrangement. We disagree with Bohr when he insists that the process is not analysable. The quantum potential provides a means of analysis.

It is when we move on to consider the many-body problem that novel features appear in the quantum potential. For simplicity, let us consider the two-body problem. If we consider a pair of particles that have a wave function which can only be written in the form $\psi(\mathbf{r}_1,\mathbf{r}_2,t)$. Then we find that the quantum potential is *non-local*. Furthermore, a closer examination reveals that as the distance between the particles becomes very large, the quantum force between the particles need not decrease. Indeed, even in the limit when the particle separation tends to infinity, the interaction need not become negligible. In other words, even if the particles are separated by many miles and there is no classical potential between them, the particles remain 'locked' together by the quantum potential. They are 'together yet apart'. Thus, as one particles moves along its trajectory, the second particle responds immediately to any change. In a way, it is something like action-at-a-distance. But action-at-a-distance is the last thing physicists want! Since Newton's time, all theories have tended to confirm locality and even Newton himself regarded action-at-a-distance as a "philosophical absurdity",[7] so surely there must be something wrong with the quantum potential approach.

But Einstein had already noticed that there was a problem when one looked at a two-particle system described by a non-product wave function, even in the usual formalism. In his paper with Podolsky and Rosen (EPR),[8] he argued that the formalism seemed to be incomplete. Let me try to explain his reasoning. First, let us consider a measurement performed on a single system. As the system is passed through the measuring device, one of the variables of the system takes on particular value. This is called the eigenvalue of the operator associated with the measuring instrument. The complementary variables become or remain undefined. It is this process that provides an explanation of the uncertainty principle. In effect, a system cannot have *all* its dynamical variables well defined, only a subset is well defined; the rest, the complementary set, remain unspecified. Let us now extend this argument to the two-particle system. When the two particles are spatially well separated and not interacting through a classical potential, a measurement of a variable on particle 1 enables you to specify the value of the corresponding variable of particle 2. Alternatively, you could measure a complementary variable on particle 1 and you can again specify the value of the corresponding variable on particle 2. But particle 2 is not disturbed in any way, as it is far removed from the apparatus being used to measure particle 1 and, because of this, you seem able to specify sharp values for all variables of particle 2. This seems to contradict the uncertainty principle, and therefore, EPR argued, quantum mechanics was not complete. Later, when he saw the explanation provided by the quantum potential, he rejected it as well, saying that he was not willing to contemplate any theory that contained a "spooky action-at-a-distance".[9] For Einstein, then, reality involved local physical processes in space-time, thus agreeing with the traditional view.

Was he right on this point? Could it be that the quantum potential approach is just plain wrong and Einstein was looking at the usual formalism in the wrong way? It was at this stage of the argument that John Bell[10] made an important contribution. By taking a two-particle system with dichotomic variables supplemented by additional variables to take care of all elements of reality and choosing a simple and reasonable criterion for locality, he was able to show that a set of four correlation functions must satisfy a certain inequality. The results are quite general and do not depend upon any specific model, provided it is local. If we calculate these correlation functions from quantum mechanics we find that the inequalities can be violated. The result then is that quantum mechanics does not always satisfy Bell's locality condition.

But what does experiment say? A whole series of experiments over the last 10 years, each adding new refinements have now shown that the quantum mechanical calculations are correct for separation distances of up to 24m (refs 11, 12). Thus, the experimental correlation functions cannot be accounted for by any local theory. Although questions have been raised about whether Bell's locality conditions is too strong or whether the experiments actually measure what they claim to measure, there is a general reluctant acceptance that nature contains some form of quantum non-locality.

If we extend these arguments to a many-body wave function and consider the extreme case when the wave function of the whole universe[13] is involved, then surely this would imply that everything is locked together—there is no freedom—no independence. But the world around us seems to be dominated by independence. How, then, do we account for separability? Again, the answer is very simple. If the wave function of, say, a two-body system can be written in the form

$$\psi(\mathbf{r}_1,\mathbf{r}_2,t) = \phi_1(\mathbf{r}_1 t)\,\phi_2(\mathbf{r}_2,t) \tag{4}$$

it is very easy to show that the total quantum potential becomes a sum of independent quantum potentials, each acting only on one particle, that is,

$$Q(\mathbf{r}_1\mathbf{r}_2,t) = Q_1(\mathbf{r}_1 t) + Q_2(\mathbf{r}_2 t) \tag{5}$$

The systems are then independent. But a product wave function is a special case of a non-product wave function. In other words, separability is now a contingent feature of nature. It is a special case of non-separability. Thus quantum phenomena seems to imply that complexity at the quantum level cannot be accounted for through simplicity. But rather it is the other way round!

Let us take a closer look at this quantum non-locality and see what we can learn about it from the quantum potential. First, I must eliminate a common worry, namely, that this non-locality, when considered in the context of relativity, will somehow lead to very profound contradictions with our normal experience. What this quantum non-locality may suggest is that events can be connected over space-like distances and through these connections we can send signals faster than the speed of light. If we can, in fact, do this, then we can invert causality by, for example, arranging to have our fathers killed before we have been conceived. Clearly, this is an undesirable effect!

Now, if the quantum potential was of the same nature as a classical potential then this would certainly be the result. However, the quantum potential is not like a classical potential. I have already pointed out some of the differences when we considered the one-body problem. But there is a further and yet more radical feature of the quantum potential that I want to bring out.[13] We know that a classical potential is always a pre-assigned function of the positions of the particles. The quantum potential depends upon the quantum state of the whole system. Change the quantum state and you change the quantum potential. This means that two sets of particles with the same set of initial conditions (that is, positions and momenta) will evolve in different ways if the wave functions of the two sets of systems are different. In other words, the quantum potential cannot be a pre-assigned function of position.

The above situation can be summarized as follows: with pre-assigned classical potentials, the parts are organized *into* a whole but with the quantum potential the parts are organized *by* the whole. In other words, in quantum phenomena the *whole has an independent and prior significance*. So that the quantum potential leads to the same conclusion that Bohr arrived at from the usual formalism, namely, the quantum phenomena form an undivided whole. The major difference is that the quantum potential approach allows further analysis, a feature denied by Bohr.

What does it mean to have the parts organized by the whole? Is not some new order required? I believe there is a new order and I will try to motivate a new direction by exploring some of the clues provided by the quantum formalism itself. Feynman[14] has shown that there is another way of looking at the evolution of the wave function and that involves essentially a Huygen's construction. The wave function at $(\mathbf{r}_2 t_2)$ is obtained from the wave function at an earlier space-time point $(\mathbf{r}_1 t_1)$ through the integral

$$\psi(\mathbf{r}_2, t_2) = \int K(\mathbf{r}_1 \mathbf{r}_2 t_1 t_2)\, \psi(\mathbf{r}_1 t_1)\, d^3\mathbf{r}_1 \qquad (6)$$

Perhaps it is simpler if we represent this integral diagrammatically as shown in Figure 4. Then for a given initial time t_1, the wave function $\psi(\mathbf{r}_2 t_2)$ has 'folded' into its contributions from all parts of the surface Ω_1. Now the quantum potential at the point $(\mathbf{r}_2, t_2)$ is calculated from $\psi(\mathbf{r}_2 t_2)$, thus the quantum potential at a given space-time point contains information from all points on Ω_1. If we block out some of the points on Ω_1 with, say, an impenetrable screen, then this will modify the quantum potential at $(\mathbf{r}_2 t_2)$.

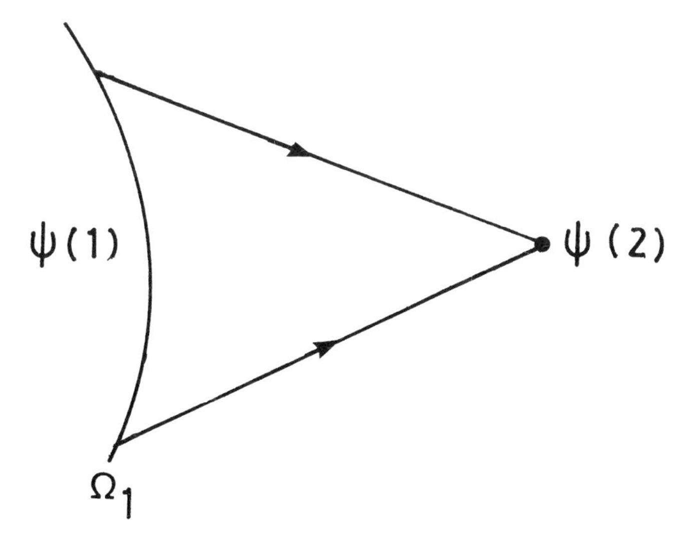

Figure 4 The diagramatic representation of the integral equation (6).

In this way, we can provide an explanation of why, in the two-slit experiment, the quantum potential in the region between the slits and the screen still contains information about the slits.

We can summarize all this by saying that each region of space-time contains information about the whole environment and the particle itself is sensitive to this information, which is carried by its field-like aspect. But this language is rather inadequate as it still contains the particle-field dichotomy. Such a division is totally out of place in a description that is based on the notion of undivided wholeness. Therefore, are we right to use the word 'particle' with all its classical connotations? Quantum field theory already tells us that particles can be created and annihilated. This is the common language in which the whole of field theory is framed. There appears to be no permanence of substance or matter. Could it be that the trajectories are simply an average of some deeper process which manifests itself with particle-like aspects? We could, say, think of a process akin to the Brownian motion of mist droplets near the critical point, namely that particle-like concentrations are always forming and dissolving. A particle-like manifestation will, of course, contain a form of inertia, so that if it dissolves at a certain point, it will re-form nearby, the whole process producing a series of concentrations from which a track can be abstracted.

There is a very nice metaphor to illustrate the whole process which has its origins in the 'unmixing' experiment.[15] Consider two plastic concentric cylinders with glycerol filling the gap between them. If a spot of suitable dye is placed in the glycerol and the inner cylinder rotated relative to the outer one, the spot of dye is drawn out into a thread and finally disappears. If the inner cylinder is now rotated in the opposite direction the spot reappears, albeit with a small diffusion. The dye has first been enfolded into the glycerol, so that its structure is latent or implicit. It can then be made explicit by unfolding it from the glycerol. A track similar to the one we mentioned above can be constructed from a series of spots of dye in the following way. Place a spot of dye at a point in the glycerol and turn the inner cylinder a few times. Place another spot of dye a centimetre behind the position of the first spot of dye and turn the inner cylinder again through a few turns. Repeat the process. If we now turn the inner cylinder backwards, we see a series of dots forming and disappearing again, giving the impression of something moving in a straight line through the glycerol. But there is no particle moving; there is only a series of spots being enfolded and unfolded, giving the illusion of a particle-like movement.

David Bohm[15] argues that this enfoldment and unfoldment is universal. There are no permanent entities, just a series of relatively-local, quasi-autonomous patterns emerging from a background which is called the holomovement. These patterns are maintained by a constant movement of enfoldment and unfoldment. That which is unfolded is called the *explicate* order, while that which is enfolded is called the *implicate* order. The evolution of the explicate order is organized by the information contained in the implicate order. It is in a scheme like this that we can begin to see how the parts are organized by the whole.

There is one final point to be made and that is that this holomovement is not in space-time. It is the relationship between features in the explicate order that determines space-time properties. Hence space-time and locality are no longer primitive but derived. As John Wheeler[16] so eloquently puts it in terms of the creation.

Day one: the quantum principle
Day two: geometry

My conclusion is that simplicity is a special case of complexity.

Acknowledgment

I am deeply indebted to David Bohm for many discussions that have provided a clarification of the radical changes that he considers necessary to provide a deep understanding of reality.

References

1. Gell-Mann, M., *The Nature of Matter. Wolfson College Lectures, 1980*, Clarendon Press, Oxford (1981).
2. Miller, W.A., and Wheeler, J.A., in Kamefuchi, S. (Editor), *Proc. Int. Symp. Foundations of Quantum Mechanics in the Light of New Technology*, Physical Society of Japan (1984).
3. Bohr, N., *Atomic Physics and Human Knowledge*, p. 64, Science Editions, Inc., New York (1961).
4. Bohr, N., *Atomic Physics and Human Knowledge*, p. 62, Science Editions, Inc., New York (1961).
5. Bohm, D., and Hiley, B.J., *Foundn Phys.* **12**, 1001 (1982).
6. Philippidis, C., Dewdney, C., and Hiley, B.J., *Nuovo Cimento, 52B*, 15 (1979).
7. Thayer, H.J., *Newton's Philosophy of Nature*, p. 54, Hafner, New York (1953).
8. Einstein, A., Podolsky, B., and Rosen, N., *Phys. Rev., 47*, 777 (1935).
9. Einstein, A. in Born, M. (Editor), *The Born–Einstein Letters*, p. 158, Macmillan, London (1971).
10. Bell, J.S., *Physics, 1*, 195 (1964).
11. Aspect, A., Dalibard, J., and Roger, G., *Phys. Rev. Lett., 49*, 1804 (1982).
12. Butt, D.K., Personal communication.
13. Bohm, D., and Hiley, B.J., *Foundn Phys., 5*, 93 (1975).
14. Feynmann, R.P., *Rev. Mod. Phys., 20*, 367 (1948).
15. Bohm, D., *Wholeness and the Implicate Order*, Routledge and Kegan Paul, London (1980).
16. Patton, C.M., and Wheeler, J.A., in Isham, C., Penrose, R., and Sciama, D., (Editors), *Quantum Gravity*, Clarendon Press, Oxford (1975).

Dr. Alan Gauld

Alan Gauld is Senior Lecturer in Psychology at the University of Nottingham.

After reading History and Natural Sciences at Cambridge, he became a King George VI Memorial Fellow at Harvard University, and a Research Fellow at Emmanuel College, Cambridge, leading to PhD in 1966.

His publications include four books on psychology and parapsychology, and his current research interests include hypnotism and philosophical problems in cognitive psychology.

He has been a member of the Society for Psychical Research since 1954, and member of the Council since 1962. He is also a member of the British Astronomical Association and the Classical Association.

Ghosts in the machine

It is a commonplace of the history of science that significant discoveries and important conceptual advances have often come from investigation of or reflection upon phenomena—sometimes quite trivial—which are out of step with current theories or assumptions. This being so, you would think that scientists as a class would be particularly interested in reports of anomalous phenomena, phenomena which do not 'fit in', and that they would lose no time in trying to verify or to discredit the observations concerned. Some scientists in some areas do have this sort of attitude. Perhaps astronomy and the cosmological sciences provide the prime examples. Cosmological theories are commonly of a highly speculative character, and may at any time be strengthened, undermined, or thrown back into the melting pot by unexpected or anomalous observations. In fact, you might almost say that in this area the unexpected is to be expected and anomalies are the rule. The database grows more complex and theories have to be reshaped accordingly. But this must be very frustrating for the theoreticians. Perhaps so; but I have a rather strong impression that most of them would not have it otherwise. They *like* a universe which may spring a surprise on them at any moment, and which presents them with a vista of possible problems and complexities. They would *not* like to find themselves once again in the situation which obtained in the days of Laplace when for a while it really seemed as though the sort of theory presented in the *Mécanique céleste* would soon give us the last word on all cosmological problems. A refractory universe, one which always is, or which soon becomes, untidy with respect to our latest and best theoretical endeavours, is so much more stimulating and challenging than one in which all theoretical problems have been finally laid to rest.

There are, however, other scientists—probably a greater number—who do not share this attitude. Their liking is for orderliness and certainty, for a solid and relatively unchanging body of knowledge which gives them, perhaps, a certain control over events, but, more than that, gives them the comfortable feeling that through their own hard intellectual exertions they have clambered to an impregnable little plateau from which they can view

Figure 1 The phantom butler seated on the left.

their world with a good measure of understanding and with no danger of having to embark on further slippery and exhausting climbs. Such 'tidy-minded' persons do not welcome anything which suggests that the world may be fundamentally more complex and difficult to understand than currently received wisdom admits.

Especially unpopular with these tidy-minded persons are reports of alleged 'psychical' or 'parapsychological' phenomena. By these terms I mean phenomena which, purportedly, involve, or constitute, either the acquisition of factual information without the use of the known sense organs (this is generally called ESP or extrasensory perception), or the direct effect of mind upon matter beyond the effective sphere of influence of the subject's body (generally called PK, or psychokinesis). Such phenomena, or alleged phenomena, have been investigated in only a few university departments, especially in America. The investigations most commonly take the form of laboratory experiments using microcomputers and associated electronic hardware, together with the experimental methods and conceptual furnishings of cognitive psychology. However, here I shall eschew this heavy stuff, and instead touch on two categories of what may be called 'spontaneous' parapsychological phenomena, namely apparitions and poltergeists. Investigating cases of these two kinds has been a hobby of mine for many years, and I must emphasize that, whatever be the ultimate explanation of the happenings in question, we are dealing not with gossip, legends or newspaper stories, but with the signed first-hand reports of seemingly reputable eyewitnesses, who have submitted to interview and interrogation. I have collected a good deal of such testimony in person; there is much, much more in the files and the publications of the Society for Psychical Research and the American Society for Psychical Research, two quite reputable associations; and, in some instances, I have witnessed the phenomena.

Surveys old and new suggest that something like one person in 10 of the general population will at some time or another have, while awake, sober, not feverish, drugged, and so on, an experience of seeing a person of hearing a voice, or both, without these experiences having, so far as could be determined, any external physical cause. Such 'visual apparitions' (I am, for simplicity, neglecting the auditory cases) usually look and behave very much like ordinary human beings, except that after a varying but usually brief lapse of time they vanish into thin air, or sometimes through a wall in the traditional manner. For these and other reasons they have generally been regarded as hallucinations—in fact, in several cases an apparition has been seen to open a door which was afterwards found to be still closed and bolted, so that not just the apparition, but the opening of the door, was part of the same hallucinatory episode. Indeed, most apparitions might reasonably be looked upon as peculiar and usually one-

Figure 2 A ghostly monk, or a double exposure?

off hallucinations of unknown cause, undergone by persons so far as is known completely sane. For example I once got the statement of a perfectly ordinary undergraduate of Queens' College, Cambridge, who had while walking down Silver Street one night towards his College, the following peculiar experience. He saw in front of him, on the left-hand side, an elderly gentleman walking ahead. He followed this gentleman for at least a minute—long enough to observe that he was wearing a hat and an old-fashioned frock coat with a button in the middle of the back, and that he had silver hair curling up over his collar. The old gentleman crossed the road before coming to The Anchor public house. My friend followed him only a pace or two behind. Just before the figure reached Silver Street Bridge it suddenly vanished—right beside a bricked-up window! So far as I can discover this figure has never been seen by anyone else. Presumably an explanation of the sighting has to be looked for entirely within the percipient himself.

Of course, the hallucination theory of ghosts might be countered with various supposed examples of ghost photographs. There are quite a lot of these, but nearly all suffer from the disadvantage that the photographer was not aware of anything odd at the time when he took the photograph; only subsequently did he discover something on his negative that he could not account for. Under these circumstances it is almost impossible to rule out an ordinary explanation. Here are just two examples of such photographs:

Figure 1 was taken in the 1890s with an exposure of several minutes, during which the photographer left the room. It appears quite possible that in her absence the butler came into the room, sat down, had a quick swig from the decanter, and made off again, leaving an imprint on the film only when he was seated and stationary.

Figure 2 is taken in a house supposedly haunted by the ghost of a monk. Alas, there is every reason to suppose it is a double exposure.

Very common in recent years has been the extra passenger in the back seat of a car. Usually the photo is of someone behind the photographer and reflected in the window glass.

So does the hallucination theory hold the field? I think we must say that if visual apparitions are hallucinations, some at least of them are not just hallucinations. For in a surprisingly high percentage of cases the hallucinations might be described as in one way or another veridical. Something about the hallucination, or something in its content, corresponds with some event or state of affairs outside the hallucination in a way for which no ordinary explanation seems viable. It is as though the hallucination is a vehicle for ESP, for information not obtained through the ordinary channels of sense.

A particularly common class of veridical hallucination is that of so-called 'crisis apparitions'. These are the cases in which the percipient sees before him the figure of a person he knows, say his Aunt Jane. He is sure that it is Aunt Jane because she is wearing her flowered hat, long skirt and elastic-sided boots, and is carrying her umbrella with the duck-head handle. Indeed, he thinks it is his Aunt Jane, until she suddenly disappears. Afterwards he finds that at or about that very time she quite unexpectedly died or underwent some other nasty crisis in her affairs. Surveys suggest that in our society as many as one person in 600 is likely to have had an experience of this general kind.

The veridicality of crisis apparitions consists primarily in the coincidence between the apparition and the death; but not infrequently the apparition itself in one way or another indicates the fact, even the manner, of death, by speaking, by exhibiting a wound, and so on. In another category of veridical hallucination, the figure seen is unknown to the percipient, but afterwards he is able to identify it from a description or a photograph with a person who formerly lived in or was associated with the place where it was seen. For example, a lady who was living in what had formerly been a vicarage gave me the following account of an odd experience she had there about a year before the interview. She was playing the piano in the music room, and felt someone looking over her shoulder. She thought it was one of her sons. She turned round and saw a lady dressed in old-fashioned clothes. Her hair was parted in the middle and drawn towards the back. She was wearing a full skirt and a blouse with lace. The figure was in black and white, not in colour. The witness ran out of the room leaving the figure still there. Afterwards she told the Churchwarden who produced some old photographs from a drawer. From these the witness was able to identify the figure as that of the wife of a former vicar, a lady who had been a music teacher.

Yet another class of veridical hallucination is that of collectively perceived apparitions, apparitions which are seen in the same place at the same time by more than one person. Such cases are not uncommon. In fact in between a third and a half of cases, where there is a second person in a position to see the figure, he sees it. You also get cases in which what is apparently the same figure is independently seen in the same locality by different people at different times. In one case which I investigated the same unmistakable figure was seen separately and collectively by the six occupants of a council house on upwards of 20 occasions in a period of about three years.

Poltergeist phenomena are very different from apparitions. They are physical disturbances (especially percussive noises, and movements or projection of small objects) which centre around particular persons (often adolescents). There is no question of the phenomena being hallucinatory or in any way unreal. They can be recorded on ordinary instruments. Figure 3 shows a photograph of alleged poltergeist activity captured by a newspaper photographer and I have recorded several poltergeist noises.

The central problem that one faces in investigating a poltergeist case is, therefore, not that of whether or not certain physical events really took place, but that of whether or not they were fraudulently produced. And without doubt there have been a number of instances of egregious fraud by the young person centrally involved—I have come across several. However, in other cases the likelihood of fraud seems vanishingly small. For example in one well-known case, the Miami poltergeist of 1966–67, two competent observers, the late Professor J.G. Pratt and Mr. W.G. Roll, both of whom I knew personally, were able to sit at either end of the principal scene of events, a large warehouse room, and dictate into tape-recorders not accounts of the ongoing phenomena as such, but information as to where each possible suspect, including the principal one, was and what he was doing at the time of each occurrence. The occurrences were principally the throwing and breaking of small china, glass and porcelain objects. In many instances, it was quite impossible to suppose that anyone had caused the disturbances by ordinary means.

Even more extraordinary phenomena have sometimes been reported, for instance objects appearing in or disappearing from enclosed spaces, or travelling through the air as though carried rather than thrown. Attempts have been made to relate poltergeist phenomena to psychological problems in the individuals round whom they centre. The idea is that repressed motives and emotions may find expression in poltergeist phenomena rather as they are supposed to in, for instance, hysterical conversion symptoms. Certainly the happenings are not usually under the conscious control of the focal person, but in recent years a few remarkable people (Kulagina, Vinogradova, Felicia Parise) have apparently been able to produce poltergeist-like phenomena before outside observers more or less at will. One of them was a Russian lady named Nelya Kulagina who has been visited by a number of Western observers. Her staple performance is to move, without touching them, though with signs of great physical stress, small objects placed in front of her on an ordinary small plastic hotel table.

Are these all just conjuring tricks? Certainly when, as in one film, you see her moving her hands above a large compass, brought by a visitor, and making the compass needle rotate, you might well think so. But a little later on in the same film she rotates the body of the compass round the needle, which remains more or less still. Two western observers brought her a small plexiglass cube, open on one side, inside which was a table tennis ball suspended by a spring. The table tennis ball was moved in various ways, and then, so to speak, held for some seconds on the bottom

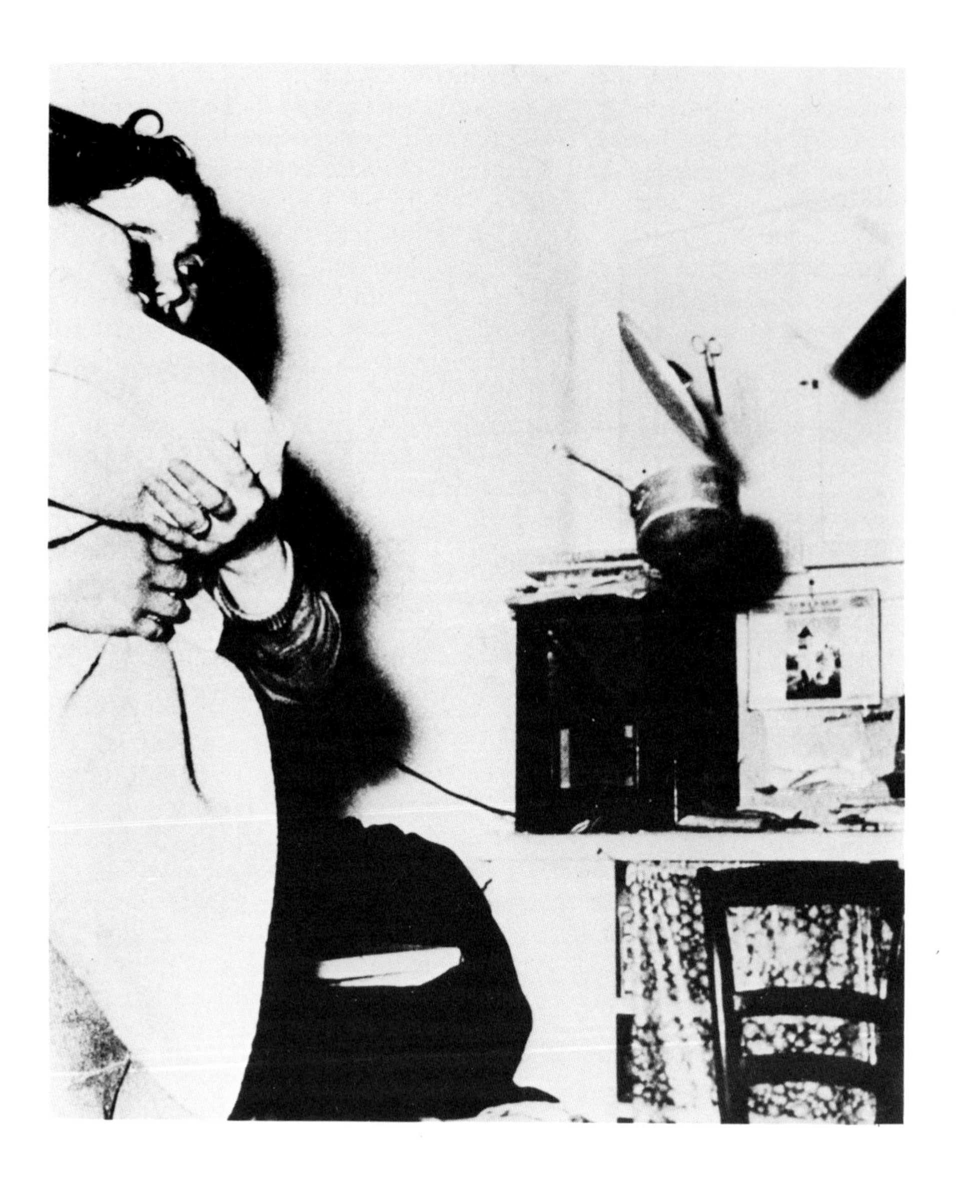

Figure 3 Alleged poltergeist activity.

of the box before being released.

Reports of such phenomena are unpopular with tidy-minded scientists for some very obvious reasons. These phenomena suggest that something is in us or that we are in something which is not subject to the physical and spatiotemporal restrictions that are ordinarily thought to limit our spheres of cognition and influence. Biologists and psychologists are particularly reluctant to believe in such transcendant faculties because, first, any attempt to explain them seems likely to demand notions far more complex and abstruse than those of the currently accepted conceptual frameworks of biochemistry and neurophysiology, and, second, it is difficult to envisage how, and from what, such faculties could have evolved. Hence the tidy-minded, holding as they do the sort of attitudes I outlined at the beginning, would for the most part very much like to dispose of all these alleged instances of 'paranormal' phenomena.

Their simplest tactic, and perhaps their only possible tactic, might be to impugn the testimony which apparently favours the genuineness of the phenomena. Many and painful researches by many and painstaking psychologists have shown us that eyewitness testimony is in many circumstances not to be relied upon. And surely it will be at its least reliable when it concerns such bizarre and exciting events as apparitions and poltergeists. Hence, we may justifiably dismiss all reports of these phenomena, and settle down again with a comfortable sigh into a Universe that does not contain too many refractory loose ends.

I do not think that this tactic is satisfactory. Looking at the body of testimony for veridical apparitions and for poltergeists, I can see no grounds at all for dismissing it, whilst at the same time not dismissing the testimony for many widely accepted phenomena which are, or were at the outset, much less well-evidenced, for example, meteorites, ball lightning, transient lunar phenomena, the existence of the giant squid. In effect, the tidy-minded are saying of 'paranormal' phenomena: You would need pretty good testimony to establish the genuineness of such unlikely phenomena; and *this* testimony *cannot* be good, because you simply cannot take the word of people who say they have seen such things. This kind of attitude is called out in the tidy-minded by many things other than reports of apparitions and poltergeists. It may, indeed, sometimes or often save those who hold it from embarrassment and ridicule; but it is also in the long run totally stultifying. However desirable it may be from many points of view to keep one's theories as simple and as elegant as is compatible with the facts, this simplicity cannot usefully be achieved by hiding from oneself the full complexity which the facts reveal. And my own view (which is, of course, almost certainly not a majority view) is that we have here facts the proper explanation of which will in the end require theories whole orders of magnitude more complex than any we now possess.

Dr. Lewis M. Branscomb

Lewis Branscomb is Vice President and Chief Scientist of International Business Machines Corporation, a member of its Corporate Management Board, and is responsible for guiding the Corporation's scientific and technical programmes to ensure that they meet long-term needs. He joined IBM in May 1972, and was named a member of the Corporate Management Board in March 1983.

A research physicist, he was appointed Director of the National Bureau of Standards by the President of the United States in 1969, after a career at the Bureau since 1951.

He graduated from Duke University summa cum laude in 1945, and was awarded MS and PhD degrees in Physics by Harvard University in 1947 and 1949.

Among his affiliations, he is a member of the National Academy of Engineering, the National Academy of Sciences, and the National Academy of Public Administration in the United States. In 1979 he was appointed by President Carter to the National Science Board, and in May 1980 was elected Chairman, serving until May 1984.

He is a director of the IBM World Trade/Europe/Middle East/Africa Corporation, the Mobil Corporation, the General Foods Corporation, is a trustee of Vanderbilt University, the Carnegie Institution of Washington, and the National Geographic Society, and is a member of the Board of Overseers of Harvard University.

A time for rediscovery

Unlike the speakers at this symposium, many people are completely overwhelmed by the complexity of the world around them. The quest for simple, universal explanations for the mysteries of life has created both religions and sciences—and often some mixture of the two.

One may deride the creationists for their refusal to accept the evidence for the migration of tectonic plates, or the successful prediction of remnant microwave radiation from the big bang. But one must admit that poetic inspiration can be a guide to science, just as it can to mysticism.

Indeed, when the author of the book of Genesis tells us that God's first command was, "Let there be light", the notion that the universe had its origin in a ball of radiation seems well expressed. After all, Genesis doesn't start out with, "Let there be *rocks*," or even, "Let there be intergalactic dust".

It is hard to think deeply about science without grappling with its philosophical implications—just as it is hard to appreciate the beauty of science without a cultivated taste for beauty in other spheres.

Among professional mathematicians and physicists, an astonishing fraction are competent amateur musicians. (I do not count myself among them, although I did sing the role of professor in satirical lyricist Tom Lehrer's *The Physical Review*—written in 1950 for the final lecture of my freshman physics course at Harvard.)

A rather smaller fraction of professional scientists seem to commit themselves to paper as amateur philosophers, and I attribute that to two causes. First, the professional philosophers in the past 50 years have withdrawn so far from the mainstream of modern creative thought that scientists have difficulty making contact. And, second, a scientist indulging in philosophy can be reasonably sure of being attacked for lack of rigor from *both* sides—philosophy and science—if he indulges.

So I award the participants in this symposium high marks for courage—if not good judgment—in exploring these fascinating notions of complexity in science and its philosophical implications. But it is important to step back from the canvas occasionally, and think a little about the nature of the scientific and technological enterprise in which each of us is engaged. It may help in understanding the evolving relationship of science to the other human activities from which it draws its sustenance, and through which it brings value to society.

In my view, for example, no meaningful distinction can be drawn between industrial and academic science. The conditions and methods conducive to rapid progress in research are *universal*, and I will say more about that later—along with the role of aesthetics in *all* research progress and the potential conflict between imagination and disciplined judgment.

I will show you examples, from IBM laboratories, of two very sophisticated, *and* very beautiful, ways of dealing with complexity; but I will also suggest that we in the fledgling field of so-called 'computer science' are only beginning to face the truly complex issues inherent in bringing information systems to the service of mankind.

Is industrial science different?

At a conference on 'Physics in Industry' organized by the International Union of Pure and Applied Physics (IUPAP) in Ireland in 1976,[1] I was asked to discuss the problems—or, as we in IBM like to call them, 'opportunities'—of a working physicist in industry.

My contention was that there is no such thing as 'industrial physics', as if we were to imagine that there is some other kind of physics with which it may be compared. The God who created physics had only one variety in mind, and those of us who seek to make out that grand design must avoid imposing contrary notions.

Of course, there is a 'physics in industry', and there are 'industrially-employed physicists' with motivations and research attitudes that may, indeed, distinguish us from our academic peers. But the differences are usually exaggerated in the minds of students and the general public.

The key elements of success in *both* industrial and academic research, are:

(1) Self-motivation
(2) Mastery of one's science
(3) Critical judgment
(4) A sense of history and of context for one's work.

Progress through aesthetic choices

The aesthetic dimensions of the scientific experience are extremely important, in industry, as well as in academic research. Appreciation for the beauty of science is the strongest source of self-motivation; the respect of one's peers is second; pay packets and titles rank third.

Given the complexity of nature as we observe it empirically, it is deeply moving to experience the predictive power of science. The extent of that predictive power guides us in our search for the right questions.

The progress of science is not marked by *answers*; they are only milestones to tell us if we are still on the right road. Science progresses by reformulating our models so that we can define fruitful new topics for investigation. In that sense, the progress of science is measured by the growing extent of our ignorance, the number of well-structured *questions* to be explored, not by the number of answers we can transfer to engineering handbooks.

Many scientists are self-conscious about discussing the aesthetic element of scientific judgment. The closest most of us come to articulating this notion are references to 'good taste' in the choice of problems, and 'elegance' in theories that describe them. These are significant notions; they are an important supplement to the notion of 'truth' derived from empirical descriptions or deductive logic.

The guiding principle for scientific truth is the maximization of predictive power with the smallest possible number of independent, arbitrary assumptions. Thus our goal is to achieve the ultimate *simplicity* in our arbitrary postulates, capable of describing the greatest level of *complexity* in nature. This goal has certain practical rewards, from an engineering point of view, but it is fundamentally an aesthetic choice.

Imagination versus objectivity

Imaginative, intuitive insights must, of course, be subjected to the discipline of critical judgment and experiment. Self-deception is an ever-present danger—particularly when self-motivation based on the beauty of nature comes into conflict with external motivation to solve a predefined problem. Wishful thinking can deceive us into believing that our apparatus, or our equations, are in a constructive conspiracy with us to reach the desired objective.

After all, 'God loves the noise as much as he does the signal.' In fact, I am sure God does not see any difference. A clue to another fundamental question may as well be buried in the noise, or in the instrument calibration runs, as in the channel of measurement we choose to call the 'signal'.

Perhaps the best demonstration of that was the discovery, by Arno Penzias and Robert Wilson of Bell Laboratories, of the remnant intergalactic microwave signal from the original cosmic explosion. While Bob Dicke and his students at Princeton were launching a search for such a signal, Penzias and Wilson stumbled on it while calibrating a microwave antenna; they had their minds, as well as their eyes, open.

Rewards of originality and aesthetics

Scientists in industry are likely to feel particularly keenly the conflicting incentives to imaginative free-thinking and to disciplined, critical judgment. For, if they succumb to wishful thinking, they are more likely to be

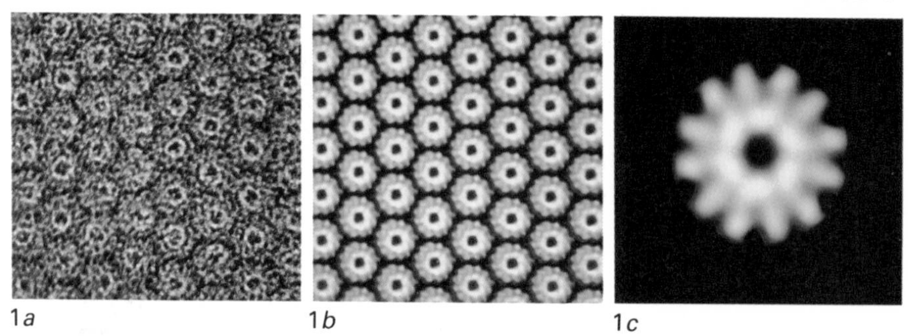

Figure 1 Fourier-based enhancement of an electron micrograph of a virus called bacteriophage phi-29 takes advantage of its symmetrical shape. *a*, Original image of a specially prepared 'crystal' of phi-29 neck slices which have been isolated from the complete virus. *b*, Same image after translational filtering, which retains only the relevant (periodic) information. *c*, Further rotational filtering of a single extracted neck reveals six inner units of protein and 12 outer units in the 'arms'.

found out by their engineering colleagues than are academics protected from immediate judgment by the low-pass filter of a pedantic literature. Yet, their imagination is rewarded by the power of a new idea to obsolete prior notions of how things work, and what might be possible.

The relationship between scientific imagination and institutional values is highly non-linear. It carries an irreversible 'arrow of time' in the social affairs of man. Original ideas can change the world forever.

The notion of provable correctness of computer programs, pursued with great success by Dr Shmuel Winograd at IBM's Yorktown laboratory, and the concept of relational data structures and the mathematical formulation of practical tools for their use, invented by Dr Ted Codd at IBM's San Jose laboratory, are two examples. But, usually, such insights come in smaller, although still very valuable, packages.

Almost always, the achievements are shared with others, some known, some working completely independently. Often, they attract the attention needed to give them practical value because of aesthetic qualities that can be appreciated by the non-specialist—not the least important of whom are business executives.

Thus, aesthetics plays a role not only in motivating the scientist, but potential patrons and users, as well. Let me cite two such examples.

Viral image processing

The first is a molecular biology project at IBM's Madrid Scientific Center which is working with Jose Carrascosa of the Spanish Scientific Research Council on the structural study of viruses.[2–5] Here, advanced image processing applied to transmission electron microscope pictures of one virus—bacteriophage phi-29—has made it possible to reconstruct the three-

2*a*

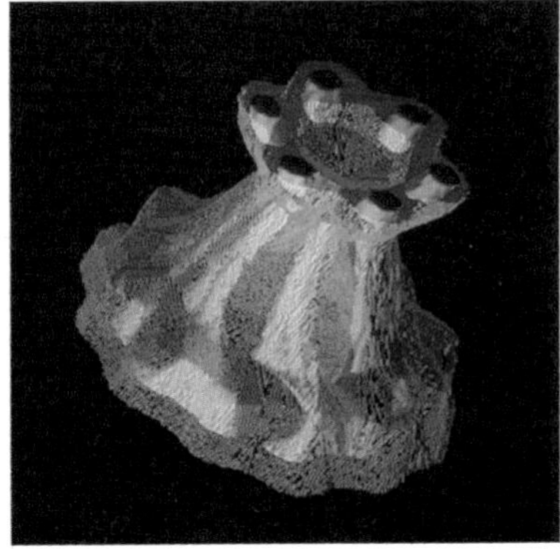

2*b*

Figure 2 Three-dimensional models of the head-to-tail connecting region of virus phi-29, about 10 nanometers long, reconstructed by computer from processed electron micrographs taken at different angles of tilt. *a*, Sectioned model of the 'neck' through which DNA moves from the virus's protein capsid into a host cell during infection, and into the capsid during virus formation. *b*, Composite of two models: a translucent representation showing inner and outer detail, and a second view in which the density level was adjusted to emphasize the dense, twisted backbones of the virus's protein subunits.

dimensional structure of the virus's head-to-tail connecting region, and helped in understanding its function.

The starting point for image processing is the original electron micrograph (Figure 1*a*) showing a two-dimensional crystal of phage phi-29 necks. The first step in filtering the periodic structure of the necks from the non-periodic noise is to transform the image to get its digital Fourier transform matched (Figure 1*b*) to a predefined, theoretic lattice-pass filter.

In Figure 1*c*, a single neck has been extracted from the image of the filtered crystal and rotationally filtered to reveal even more detail. We begin to see the apparent six-fold symmetry of the central region, as well as the 12 external units and the central hole through which DNA moves in and out.

More exciting still is the three-dimensional reconstruction of the neck of bacteriophage phi-29 shown in Figure 2*a*, computed from electron micrographs taken at different angles of tilt. The larger zone at the right corresponds to 12 molecules of the structural protein p10, while the smaller end shows the six molecules of protein p11. The maximum diameter of this reconstruction is 13 nanometers; its length is 10 nanometers, and the resolution here is 2 nanometers.

Figure 2*b* shows a color reconstruction of the same phi-29 neck, more computer graphics techniques have been used to show, by transparency, the most electron-dense areas of the protein molecules.

Fractal geometry

My second example of aesthetic science is 'fractal geometry', a discipline conceived and developed by IBM Fellow Benoit Mandelbrot[6] for analysing the irregular patterns that are common in nature, but beyond the power of Euclidean geometry. Benoit's colleague, Dr Richard Voss, lectured on this topic last year.[7]

Fractal forms (the word is based on the Latin root *fractus*, meaning broken or fragmented) occur naturally in a wide variety of contexts: from veins and arteries, to coastlines, to clusters of galaxies. And, with a little help from a computer, mathematical fractals can produce strikingly realistic *imitations* of nature—like 'Planetrise over Labelgraph Hill,' shown in Figure 3, which has become the 'signature' of Mandelbrot's work.

His methods are built on the ideas of several mathematicians who worked around the turn of the century on irregular, locally non-linear

Figure 3 Computer-generated 'Planetrise Over Labelgraph Hill' demonstrates the power of fractal geometry—which deals with broken or 'fractured' shapes, including surfaces so jagged they have more than two dimensions—to reproduce the irregularities typical of natural structures.

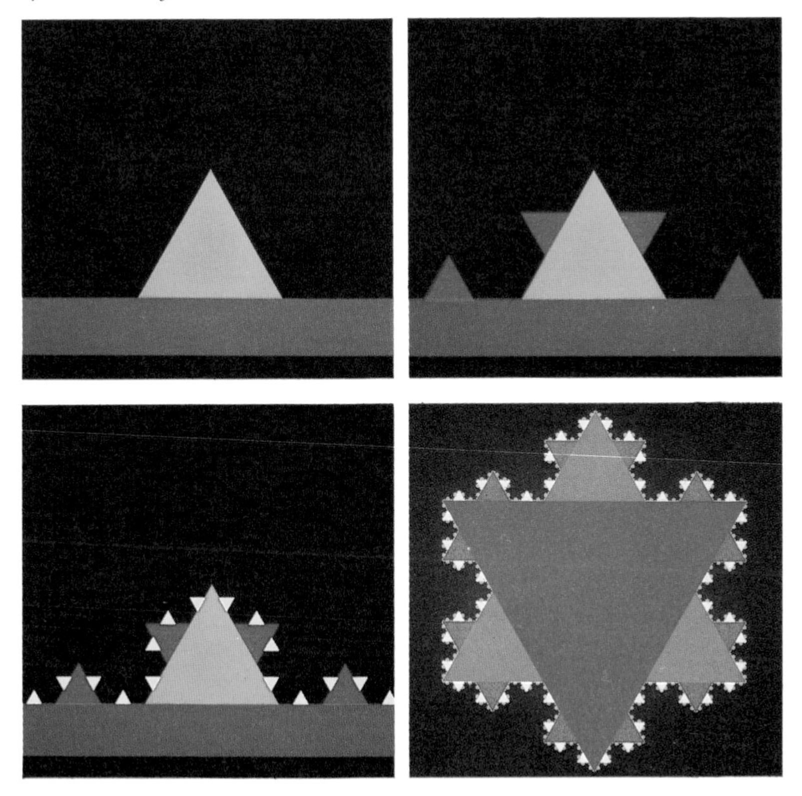

Figure 4 Koch's classic 'snowflake' curve illustrates one procedure for constructing a fractal curve: adding more and more triangles to the sides of larger triangles until, ultimately, the boundary crams an infinite (and infinitely detailed) length into a finite area of the plane, without intersecting itself.

shapes, like the classic 'Koch curve' produced by adding more and more ever-smaller triangles to the sides of larger triangles to form an infinitely intricate 'snowflake' curve (Figure 4).

The essential feature of a fractal (unlike the shapes of standard geometry, which, when enlarged, become increasingly smooth) is that the fractal's texture retains the same fine-grained irregularity, or 'wiggliness', no matter how much it is magnified. Thus, fractals suggested a simple way for computers to generate a lot of irregular detail, without the programmer having to do a lot of work.

The fractal 'landscape', shown in Figure 5*a*, for example, was produced by graphing the sums of Fourier series with Gaussian coefficients, thus simulating the unpredictable details in nature that make one mountain slightly higher or longer than another.

Figure 5 'Gaussian Hills That Never Were' takes its title from the probability distribution ruling the difference in altitude between any two prescribed points on the map. As the fractal dimension, D, of the relief increases, so does the surface roughness. *a*, $D = 2.15$; *b*, $D = 2.5$; *c*, $D = 2.8$. *d*, Scaling the height variations by a power law produces a more realistic, flat-bottomed valley.

Figure 6 Artful 'dragons' are visual manifestations of the iteration of relatively simple formulas and illustrate the basic fractal property that larger patterns are composed of smaller and smaller elements, which all have the same shape. *a*, a two-dimensional dragon. *b*, each color band represents one piece of the dragon as it grows from two to three dimensions.

Another interesting thing about fractals is that they are not limited to the whole-number dimensions of Euclidean geometry, in which lines exist in one dimension, planes in two, and so on. A fractal's 'dimension'—a measure of its roughness—can just as well be a number like 1.5 or 2.78.

As the fractal dimension of this 'landscape' is increased, (Figure 5*b*) you can see the relief of the mountains becoming more 'corrugated' until eventually (Figure 5*c*) it becomes too corrugated to be mountain-like at all In Figure 5*d*, not only has the fractal dimension been reset to a more realistic level, but to account better for the valleys (which, in nature, are smoother than mountain tops) the vertical scale has been caused to decrease with altitude.

While some fractals can easily be mistaken for photographs from nature, others, like the 'dragon' in Figure 6*a*, can be mistaken for the product of an artist's wildly fanciful imagination—perhaps a work by M.C. Escher.

The equations that generate these shapes are surprisingly simple; yet, the fractals themselves can be structures of great richness and often beauty. Indeed, a general feature of fractal geometry, which it shares with nature, is that simple formulas, repeated very many times, typically generate patterns of totally unexpected complication.

In Figure 6*b*, each color band represents one piece of the 'dragon' as it grows from two to three dimensions. It was generated by the computer plotting the dragon's two-dimensional shape on a three-dimensional grid within the machine—using an algebra called the 'quaternions' that extends the notion of 'number' to four dimensions.

Perhaps the 'grand dragon' of all fractals is shown in Figure 7*a*, now called the 'Mandelbrot set', which is a graphic representation of an infinite number of mathematically-generated fractal dragons. As this remarkable object is progressively enlarged (Figure 7*b* to 7*g*) and we view just a small area of each previous image, each tiny spot grows into a wondrous new shape, more different from the others than snowflakes. The increase in structure as one continues to zoom in is incredible. The more the Mandelbrot set is magnified, the more detail appears.

Beautiful ways to deal with complexity

These examples can usefully serve two other points.

First, their fascination owes quite a lot to the beauty of their images. In fact, fractals can not only be used to generate plausible moonscapes, but are already helping to create motion pictures like *Star Trek II*.

And, second, both are illustrations of sophisticated efforts to deal with complexity. Scientists at various institutions are attempting to use this new 'language' to shed additional light on the structure of metals (which exhibit fractal patterns in the shape of fracture surfaces, model the atomic structure of glass, study the chaos of turbulent fluid flow, and discern the patterns of vegetation from aerial photographs.

The use of relatively simple algorithms to generate complex fractals that reproduce patterns we see in nature could represent an alternative approach to generating physical theories. The traditional Newtonian method starts with a simplified model of nature, and elaborates the mathematical description in progressive steps—seeking to accommodate ever more of the real world's complexity within its predictive power.

Cellular automata are being used experimentally, not only in generating fractals but in other computer experiments in mathematics, physics and biology. So far, not much real progress has been made, but one cannot help but think that Leonardo da Vinci, who believed the eye was the critical instrument for learning, would have been an enthusiast for this approach.

Not only is the mathematical basis underlying fractal geometry more complex than the relatively simple fractal equations would suggest, but the task of understanding under what circumstances fractal distributions can be used as a model of nature is far from obvious, particularly in fields like econometrics, which has been one focus of Mandelbrot's work.

In the case of the virus images, the problem is that the algorithms used in image enhancement must be designed under the constraint of as much knowledge of the real object as possible, for example, its symmetry properties.

This really should be thought of as an application of artificial intelligence, in the sense that the process used combines, in inextricable ways,

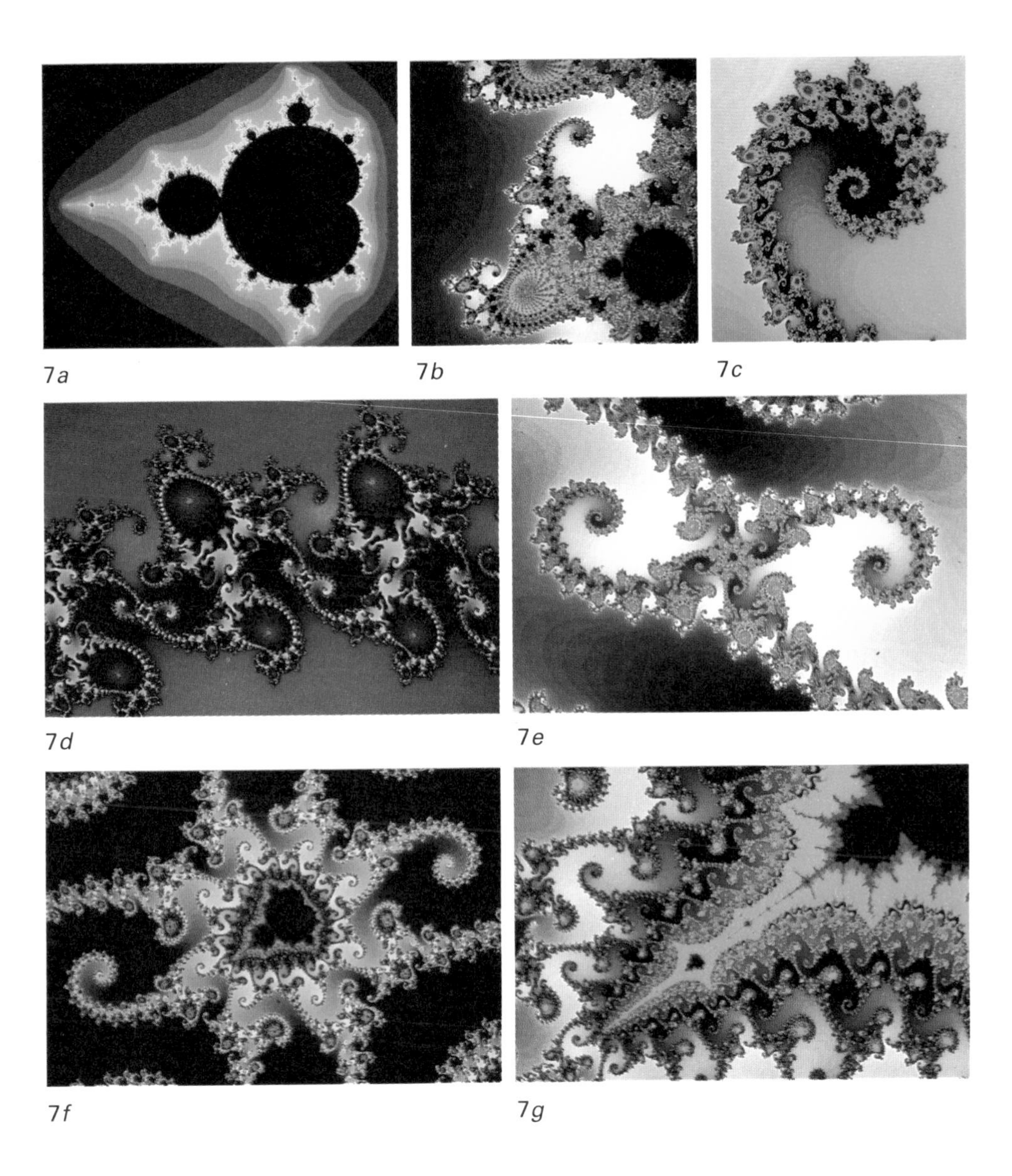

Figure 7 Keystone of fractal dragons is a shape called the Mandelbrot Set after IBM mathematician Benoit Mandelbrot who originated fractal geometry. *a–g*, as the design is magnified again and again, ever more detail appears, as in self-similar natural shapes that tend to repeat themselves (such as a tree's large branches and small twigs).

mathematical means for noise averaging and feature extraction with human judgment about the object being looked for. Does the final image discover the most important thing nature has to tell us? Or does it only reveal a lot of previously unseen detail about what we expected, or hoped, to find?

Mathematics, machines and people

These two projects are paradigms of the problems facing every computer scientist, for computer science is the only field of science built around the study of an artifact of man.

One may study the computer as an abstract device capable of a variety of mathematical manipulations. Alan Turing, indeed, laid the foundations for modern computer science by doing just that. But the real motivation in this field lies, instead, in the value of information machines to people.

To make progress, you have to know about mathematics, machines and *people*. That third element, *people*, is of overriding importance; and, yet, it is the one most often overlooked by the disciplinarians of academic computer science. Indeed, I do not know of a single doctorate curriculum in computer science in any university that requires the study of the human beings for whom the computing machines are to be designed.

The level of complexity involved in this relationship between mathematics and psychology far outstrips the complexity problems faced by the student of large departures from local thermodynamic equilibrium. Figure 8, for example, which is a computer-enhanced nuclear magnetic resonance image of the inside of my own brain, does not give a clue to the properties of a human that should guide computer science.

Interestingly, computer scientists often appeal to aesthetics to guide them, in the absence of a quantitative and predictive theory for the behaviour of information systems as perceived by humans.

Thus, for want of verified principles with which to test system and software designs, or even to judge the 'betterness' of one design over another, the information industry tends to grasp at the arbitrary and rather meaningless concept of 'user friendly'. Academics, on the other hand, tend to define an equally arbitrary artificial model of what a human might expect from a useful machine, which they call 'artificial intelligence'.

Since there are many entrepreneurs in today's academia, and since scientists in industry tend to follow the fads defined by leading universities, both groups tend to cover their bets by asserting that only with artificial intelligence techniques can one build a truly user-friendly machine.

Meanwhile, only a few brave souls such as Don Norman at the University of California-San Diego; Stewart Card and Tom Moran at Xerox PARC; Steve Boies, John Carroll, Lance Miller and John Thomas of IBM

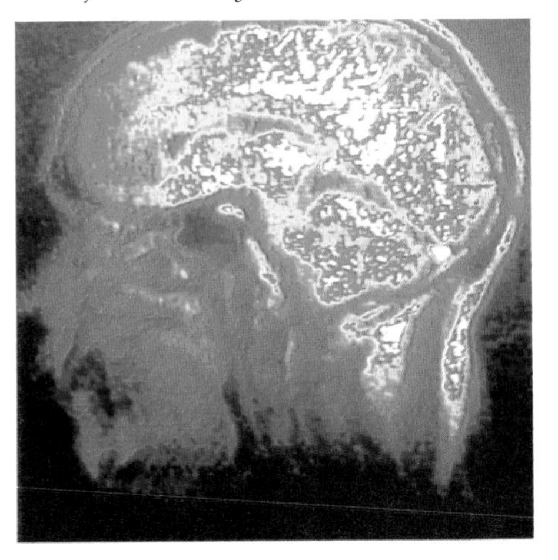

Figure 8 Color-enhanced nuclear magnetic resonance (NMR) sagittal view of a head (in this case, the author's). In NMR tomography, the combination of a magnetic field and radio frequency excitation causes atomic nuclei naturally present in human tissues to reemit characteristic signals which can be processed by a computer to yield unusually complete and detailed cross-sectional images.

Research; Philip Barnard at Cambridge University; Benedict Du Boulay at Sussex, Brian Shackle at Loughborough, and Tim O'Shea at The Open University are trying to build a rigorous behavioural science under the value system for computer science.

The right problem?

My own view is that we should focus on more than giving the machine the capability to manipulate its stored information in a more sophisticated manner—mimicking a caricature of a rational model of human reasoning which is the goal of 'expert systems' research.

We must also try to understand the behaviour that the machine must exhibit if we are to find it a congenial information assistant. There is a famous bit of doggerel written by a student on the bulletin board at the University of Wisconsin computer room:

> "I'm sick and tired of this machine;
> I wish that they would sell it.
> It never does just what I want,
> But only what I tell it."

I call this the quest for 'artificial personality', rather than 'artificial intelligence'.

Behavioural science and cognitive science must go hand in hand. Unhappily, behavioural science is not yet a science—if I define a science as a method of discovery through which false statements can be disproved. Thus, computer science is not a science, either.

I understand why mathematically-gifted persons who want to get on with research in computing define a kind of neo-computer science that measures itself against a very restricted caricature of the human beings to which the machines relate.

There are many very tough, interesting and worthwhile problems for which the limited model is quite adequate, for the artificial intelligence model of the human is endowed with human speech, hearing and sight, along with facts and inferential rules.

It is not surprising that the best work in artificial intelligence is in just these areas: linguistics, speech recognition, image processing. And it is evident that computers can be readily optimized for tasks in computation, robotic control and text manipulation, because these tasks already carry the constraints of subordination of human passion to the discipline of mathematics, machines and character representation of language.

My conclusion is that almost everybody is working on the wrong problem, or, at least, on the easy problem. Yet, that is how science progresses, step by step, probing nature to find a good question, making progress where you can. That is also the way that business progresses.

But it is still important to keep the big picture in view. As Louis Pasteur wrote, "In the fields of observation, chance favours only the mind that is prepared".

In business, to advance the notion of a philosophy to guide the research effort would be an unfamiliar choice of words, but not an unfamiliar concept. There is certainly a set of guiding principles that must motivate our research if we are to have the best chance of making progress in a revolutionary, rather than evolutionary, way.

Good taste in formulating questions is important, and so is elegance in their solutions. But most important of all is *relevance to human values*. So long as our understanding of human beings is beyond the reach of science, we must trust our sense of taste and aesthetics, and let writers and artists help guide our hands.

So, if you ask about how IBM scientists cope with complexity, the answer is, with difficulty, recognizing the limitations of our science. For the root of that complexity is *people*; they are the users of our machines.

References

1. Branscomb, L.M., "Physics and industry", in O'Mongain, E., and O'Toole, C.P. (Editors), *Physics in Industry*, pp. 1–10, Pergamon Press, Oxford (1976).
2. Carrascosa, J.L., Vinuela, E., Garcia, N., and Santisteban, A., "Structure of the head–tail connector of bacteriophage phi-29", *J. Molec. Biol., 154,* 311–324 (1982).
3. Santisteban, A., Garcia, N., Carrascosa, J.L., and Vinuela, E., "Digital image analysis of complex periodical viral structures", in von Bally, C., and Greggus, P. (Editors), *Optics of Biomedical Sciences*, pp. 48–51, Springer Verlag, Berlin (1982).
4. Carrascosa, J.L., Carazo, J.M., and Garcia, N., "Structural localisation of the proteins of the head to tail connecting region of bacteriophage phi-29", *Virology, 124,* 133–143 (1983).
5. Carazo, J.M., Santisteban, A., and Carrascosa, J.L., "Three-dimensional reconstruction of bacteriophage phi-29 neck particles at 2.2 nm resolution", *J. Molec. Biol.* (in press).
6. Mandelbrot, B., *The Fractal Geometry of Nature*, W.H. Freeman and Co., New York and Oxford (1982).
7. Voss, R., "Random fractal forgeries: from mountains to music", in Nash, S. (Editor), *Science and Uncertainty*, pp. 69–85, Science Reviews Ltd, Middlesex (1985).

Dr. W. Graham Richards

Graham Richards is an Oxford University Lecturer and Fellow of Brasenose College.

Graduating from Brasenose College in Chemistry, followed by PhD in 1964 and an ICI Research Fellowship, his career activities have included Visiting Professor at Stanford University and the University of California, Berkeley; Chairman of the Oxford University Industry Committee; Corresponding Member of the European Academy of Arts, Sciences, and Humanities; Senior Dean of Brasenose College; and a number of research consultancies.

His current research includes applications of theoretical chemistry to small molecule spectroscopic problems and drug design, using a wide range of computing facilities. He is author of about 150 scientific articles and eleven books, and adviser to Oxford University Press.

Nature unfolds—left, right and rorschach

A simple yet intriguing experiment which can be performed whenever one has a reasonably large group of people is to find out just what percentage is left-handed. The result is not 50 per cent, nor a tiny fraction but something more than one person in 20: a curious result which cries out for explanation and a problem which can be investigated at several levels of science, going ever more deeply towards fundamentals.

The facts—sociology

If we stand naked in front of a mirror then the body has an apparent left-right symmetry, yet most people favour the use of the right hand. This is only true for human beings. It is not true, for example for the higher apes. As far as we can judge, it has always been so. If we look at the cave drawings in the Dordogne in France, the evidence is that drawings of hands are left hands where the right-handed artist has traced or thrown powder on his left hand. Drawings in Egyptian tombs show right-handed warriors and hunters. What is true of the old world seems to be also true of the new: Aztec art again depicts gods as right-handed.

Even the very words are weighted in favour of right-handedness. English 'right' and 'left' are less obviously unbalanced in value than the Latin *dextra* and *sinistra* or French *droite* and *gauche*. When one comes to slang, in almost every tongue descriptions of left-handers are highly derogatory, such as 'cack-handed' referring to the supposed practice of using the left hand to wipe ones bottom: to this day in Arab countries it is not done to put the left hand into the communal dish.

The ancient religions and superstitions associated left-handedness with the devil, bad luck or witchcraft. Tarot cards, for example, depict the devil as being left-handed.

With all this prejudice it is surprising that the percentage of the sinistral minority remains so high, especially as so many aspects of daily life are made particularly awkward for them. Tools, both ancient, like scythes, and modern, like scissors, are made for right-handers. So too are cheque books and countless simple everyday objects. Musical instruments are designed for right handers and although it is possible to restring a violin so that it can be played by a left-hander, this would cause difficulties in an orchestra and is rarely done. One exception was Charlie Chaplin who played with a restrung fiddle.

Only perhaps in sport do the minority have an advantage, largely because the majority find them awkward opponents. The success of left-handed boxers is far more than statistical, although the word 'southpaw' so often used in the sport has a baseball origin. The old Chicago West Side ball park, where the expression arose, like all ball parks, is oriented so that a left-handed pitcher has his throwing arm on the south side of the stadium. In this sport too, left-handed batters are marginally closer to first base and the first baseman is generally a left-hander so that while wearing his mitt on his right hand he can cover more of the infield while remaining close to his base.

The sport of tennis is dominated by left-handers, Bjorn Borg being the exception until he left the courts to McEnroe and Navratilova. The only sport where it is illegal to play left-handed is polo. Here if left and right-handers met then the unfortunate ponies would clash head-on.

Writing is a special problem for left-handed children so that it is surprising that the ancient written languages, Hebrew, Arabic and Etruscan all go from right to left. Probably they were cut on stone or drawn with a stick in sand or mud so that the problems of smudging handwriting were not so severe. It was, not surprisingly, the sophisticated Greeks who introduced writing from left to right. First, they introduced Boustrophedon, or writing as the ox ploughs; one line right to left followed by one left to right and so on: a system reintroduced by IBM for lineprinters. They must have noticed the greater ease of left to right for the dextral majority so that the modern western form of handwriting is definitely geared to right-handers. Only occasionally do the minority have an advantage in writing. Leonardo da Vinci wrote with his left hand, possibly as a result of an accident. He could write a form of mirror writing which is possible for some left-handers and, indeed, used this as a code in his notebooks.

This small selection of the many fascinating facts about the asymmetry between left and right produce a desire to explain them. Were the origins evolutionary? Perhaps the heart being on the left is better protected by a shield held in the left hand with the sword in the right. That and many similar evolutionary suggestions have been made, but all are unconvincing and often circular.

Are the origins genetic? Two right-handed parents have only a 2 per cent chance of having a left-handed child, whereas if both parents are left-handed the chance is as much as 46 per cent. Although that suggests a simple genetic origin this is countered by the fact that in twins, both identical and non-identical some 25 per cent have one twin left-handed and one right-handed. Research has also indicated that the percentage of left-handers is increased among children who have suffered minor brain damage (although it must be emphasized that being left-handed does not imply brain damage). Facts of this type suggest a biological origin and provoke an examination of biology for explanation.

Biology

If one does stand naked in front of the mirror then if you are male there is a noticeable break in symmetry in that one testicle hangs lower than the other. This is an external example of what internally is quite common. Organs are not symmetrically placed. Our hearts are on the left in the vast majority of people. Most important, in the context of handedness, the brain is not symmetrical.

The left hemisphere of the brain controls the movements of the right hand together with things logical, language and speech. The right half, as well as controlling the left hand, seems to be more involved with pattern recognition, intuition and things artistic.

These notions have been much clarified by the split brain experiments performed for relief of intractable epilepsy where the surgeon cuts the corpus callosum preventing signals from passing from one half of the brain to the other. Patients who have undergone this operation have taken part in experiments where information is provided to only one half of the brain at a time. If, for example, the image of a spoon is directly solely to the left half of the brain then the subject immediately provides the word 'spoon'. By contrast, when the same image is presented to the right half of the brain the subject can recognize the shape but fails to provide the word, confirming that word skills are localized on the left and spatial aptitude on the right of the brain. A right-handed patient will have more difficulty drawing a perspective view of a cube with the left hand than with the right. Post operatively, however, the representation drawn with the left hand, using the right-hand side of the brain, is very much more realistic than the drawing done with the right hand.

This type of experiment has led to much speculation about the basis of differences between Eastern and Western cultures using different halves of the brain. Women too seem different from men in that skills are not quite so firmly lateralized on one side of the brain. After strokes women can often use the right side of the brain to control speech much better than can men, although the sex differences also seem to depend on the age of puberty.

Rorschach diagrams—symmetrical ink blot figures are used by some psychologists to study the brain but these do not have a clear scientific basis.

Apart from handedness and the location of organs there are many asymmetries in the animal kingdom. Among the more interesting examples, possibly related to the lateralization of speech, is a similar situation with birdsong. If the appropriate nerves which serve the two halves of the brain are cut in young birds the song develops, but the same cutting of nerves in mature birds can remove the ability to sing and shows that bird song is localized in the left hemisphere of their brains.

In plants and many invertebrate species with shells, asymmetry is visible in the form of helical patterns or coils. The Flanders and Swan song about the honeysuckle and the bindweed reminds us that one coils to the left while the other coils to the right. Most significant in scientific terms, however, are snail shells which show helical patterns which are inherited and where the protein product of a single gene may be used to alter the genetic preference.

Embryologists have noted that embryos develop asymmetrically with the left side growing first, perhaps indicating the origin of the left-dominant brain. Geschwind suggested that this lack of symmetry in development is caused by testosterone levels in the uterus, but that theory conflicts with the fact mentioned above that identical twins are often one left-handed and one right-handed.

The fact that asymmetry leading perhaps to handedness is present at the earliest stage of life prompts the attempt to peel off another layer and go deeper than biology into chemistry.

Chemistry

Conception of a new and asymmetrical being has occurred once the new entity has its own genetic material in the form of DNA. The DNA molecule is itself an asymmetric right-handed helix whose structure can be visualized with the help of models or by the use of computer graphics. The helical DNA molecule itself coils in a helical manner giving supercoils with the degree of uncoiling possibly being the method of controlling transcription. Coils of coils can again form further coils rather in the manner of ropes so that the helical patterns on the surfaces of seeds may reflect the helical nature of their consituent DNA.

The DNA codes for proteins whose structures are known in some cases from studies in X-ray crystallography and again can be visualized with the aid of models or with computer graphic displays. These show how in proteins too there are large regions of helical structure: always right-handed helices.

The helices in proteins are inevitably of one handedness because the polymers are built from monomer building blocks of just one unique asymmetric form. The constituent amino acid are of just one mirror image type. Indeed, all life is built from amino acids of one of the two alternative asymmetric forms.

This fact was recognized first by Pasteur in his historic studies of tartaric acid and is still used as the basis for the search for life on other planets, and in seeking evidence for Hoyle's notions that living forms may have been brought to Earth on meteorites or coments.

Chemists recognize the difference between left- and right-handed forms of molecules by their action in rotating the plane of polarized light. Other properties such as chemical properties are identical. Since the two forms of an asymmetric molecule appear to be identical in energy, the question which begs to be answered is "why in living systems are all the building blocks of one specific handedness?".

There have again been many speculations. One possible idea is that all living systems can be traced back to one single molecule—the 'Adam molecule', which just happened to be of one type rather than the other. Although this idea appeals to physicists it is not popular with chemists who see the behaviour of molecules in a more statistical way. If the conditions on Earth were favourable to the production of a molecule in one place, it seems unreasonable to suppose that conditions were not ripe for many more at the same place and in different locations.

An alternative range of hypotheses postulate a prebiotic imbalance between the molecules of different handedness but otherwise identical properties. There are many natural inorganic asymmetric systems, such as quartz or clay. The problem with these theories is that any mechanism which produces left-handed molecules will also exist for right-handed molecules. What is needed is some in-built asymmetry. Among the suggestions that have been made along these lines is the preferential destruction of molecules of one type by sunlight which is partially polarized by reflection from the sea. Experiment has not supported this hypothesis, any more than the suggestion that breakdown by radioactive decay could damage one type of molecule more than its mirror image if the radiation were polarized. Failure to produce a convincing chemical theory has forced attention to be focused one level deeper to enquire whether any asymmetry in the basic laws of physics can result in the molecules of life being preferentially of one optical form.

Physics

As discussed by Paul Davies, the weak nuclear force can cause an asymmetry in interactions between electrons and the nucleus which break parity. To some people, this has left open the question as to whether the imbalance between right- and left-handed molecules in nature would result from radiolysis following β-decay, but as has been mentioned this has little experimental support.

Of the several experiments which demonstrate the loss of parity and a preference for left or right in what appears to be a totally symmetric situation, perhaps the clearest and simplest is the example of the rotation of polarized light by atomic bismuth. The experiment parallels the simple arrangement which demonstrates that one optical isomer of an asymmetric molecule will rotate the plane of polarization of light. This is simple to comprehend when the active medium which rotates the plane of polarization is a collection of molecules all of which can be described either as left-handed or as right-handed. The spherically-symmetric bismuth atoms achieve a rotation of polarized light due to the asymmetry of the weak interaction although the magnitude of this rotation is minute (10^{-5} degrees in one published experiment).

In principle, this asymmetry of the weak interaction could yield a very small energy difference between the two possible mirror image forms of molecules. The minute differences between pairs of molecules could then possibly be amplified by polymerization of the individual molecules into long chains. Even greater amplification would be achieved if this multiplication was three-dimensional rather than one-dimensional. Speculation about the 'original' asymmetric form from which the symmetric molecules of life evolved are rife. One of the more satisfying ideas which conforms with the view that a three-dimensional structure would best amplify minute energy differences between mirror forms of molecule is that of Cairns-Smith. He postulates a 'low-technology' inorganic replicating system which would have predated an organic version such as the nucleic acids used by contemporary life forms.

The main alternative hypothesis for the origin of handedness in the molecules of nature is due to Prigogine who gives an account of his work elsewhere in this book. His ideas are based on non-equilibrium thermodynamics, but can give rise to two alternative interpretations.

The two forms of a molecule, the left- and right-handed pair, could have the same energy at equilibrium but when forced away from equilibrium there may be some loss of symmetry which forces us to obtain just one form as the product of a reaction. This is the view supported by Prigogine who has shown that it is mathematically possible, but whether physically realisable is open to question. Under this hypothesis, if the experiment of the act of creation were to be repeated time and time again we would always obtain the same asymmetric result: always say left-handed molecules.

A variant, also based on Prigogine's work, is the notion that we may start with two molecular types (left- and right-handed) which have equal energy at equilibrium. When forced away from equilibrium, however, chance gives us a product of one handedness rather than the other. Under this regime, if we were able to recreate the experiment of creation then we would always end up with molecular products of one mirror form but there would be a 50 per cent chance on each occasion that the product was the one described as left or as right. Amongst these possibilities about the origins of a world based on one asymmetric type of molecule, no one at present is sure which if any represents the truth.

Conclusion

This story has been unashamedly reductionist, but have we explained complexity with simplicity? The answer is certainly *NO*. There is still no clear answer to the question of why we have life of one dominant handedness. In many ways we still have not progressed very far with the question since the time of Blake who posed this problem poetically as

> "Tyger! tyger! burning bright
> in the forests of the night.
> What immortal hand or eye
> Dare frame thy awful symmetry?"

Professor David J. Evans

David Evans is Professor of Computing and Head of the Department of Computer Studies at the University of Technology, Loughborough. He graduated in Pure and Applied Mathematics at the University College of Wales, Aberystwyth.

After National Service in the Royal Air Force servicing telecommunication and radar equipment, and a short period as an Aerodynamicist with Rolls-Royce, he went to Southampton University as a Research Assistant to work on Jet Engine Noise research, for which he was awarded the MSc (Eng) degree. A further period with Rolls-Royce, in the newly-formed Digital Computer Department, brought him into contact with the early IBM computers—the CPC, 650, and 704s, on which a variety of complex engineering problems were solved.

His appointment as Research Fellow in the Computing Machine Laboratory of the University of Manchester led to work with the early British computers, and as a member of the design team of the Atlas computer he had his first exposure to Parallelism in Computer Design, which earned him his PhD degree. A term as Director to the Computing Laboratory at Sheffield University ensued, followed by his present appointment at Loughborough. He is the author of seven books and over 200 scientific papers, for which he was awarded the DSc (Wales) degree.

Parallelism in computer systems: from revelation to realisation

---

In this lecture, I want to explore the theme that *complexity is achieved from many simple ideas interacting* in our future computer systems: *parallel computers*. These computers can carry out more than one operation at the same time, that is, several independent processes each processing its own sequential instruction scheme. The motivation to work on or to develop these machines is that the performance would be increased by having more processors or the same power for less cost. This gives greater reliability because if one processor fails then there are a few more processors in support provided the system can readjust itself. We can have more system facilities available, hopefully simpler to use, and we can expand upwards once we develop the technology of making all these machines work together. But where will we use these parallel machines? They will certainly increase the speed of computation and solve problems that are too large or too time consuming for serial computers, such as the weather and other real-time applications, as well as on-line applications in speech processing, image processing and so on.

Although the speed of devices has increased since the computers from the 1950s which all had some mild forms of parallelism, and this, combined with improvements in the algorithms, increased the performance in our machines. However, we are reaching a limit and the next stage is to go into parallelism on a large scale.

There are four general levels in which system performance can be improved. We can make our devices and hardware faster, by improving the system architecture, we can improve the system organization, and we can improve the system software. The final level to be considered is parallelism.

This should produce greater scope for flexibility, availability, reliability and, of course, it must be done on a cost effective basis.

In the past, we have achieved concurrent execution of several different programs. The size of the memory of our computers was expanded in the middle 1960s then, we learnt how to build large memories, and instead of having one job inside the machine, we had several, and thus we could accomplish multi-programming. However, on many occasions when we accessed the memory we would be accessing related words and have to wait until we could access the next word, and so we started building interleaved memories. Even earlier, by overlapping input–output activities the CPU was released to work on its own. We then had separate input–output processes to do simultaneous input–output operations. Also we considered various things like pipe-lining. An instruction inside the computer consists of many stages, for instance, you have to decode the operation code, you have to call up the operands, and then you have to store the answers. In fact, during this pipe-lining stage, we learnt that while we were decoding the operation code of one instruction, we could be actually storing the answer from the previous instruction. More recently, we have now learnt how to build memories where we can do associative or content addressable storing.

Let us now consider some of these early ideas. Some of the early computers in the 1950s, possessed serial adders where only one bit was added at a time. This was a slow process but the circuit was very simple, and only one adder was needed. To speed it up, we developed the first parallel adder in which there are more circuits involved that allow all the inputs to be presented at the same time instead of sequentially. Then we went on to develop input–output channels because the CPU was linked directly to the input and output devices and could only do one thing at a time, so that any I/O bound job meant that the CPU was standing idle. Once we put in these channels we released the CPU and had an abundance of extra power. These were some of the early ideas of concurrency which is very closely related to parallelism.

We are now doing multi-processing, which are the parallel activities at the process level. Some of the recent compilers now used, optimizing compilers, actually take a piece of code unroll the loop, and then find out whether there are any independent integers in it, to see whether they could be done in parallel. So here we are looking at parallelism at all stages of our computation process.

The new multi-processor systems can be defined as 'where the number of processors are greater than one, and the main memory is available to all the processors'. So as these processors access the memory they will clash, and in fact, degrade the performance of the machine. It is surprising how few processors are necessary to get a very poor performance.

Our Von Neumann machine is called an SISD (Single Instruction stream Single Data stream) and has one memory, one processor, one control unit. We can then consider the SIMD design (Single Instruction stream Multiple Data stream). This machine has many processors obeying the same instructions via one control unit. In other words, it is a synchronous machine. To some extent they are easier to program and to build as well. Some array machines fall into this category. The best analogy of this design is that if you are the owner of 20 dogs, and you decide to take them for a walk. Now, if you put them all on to the same lead, you can manage to proceed because as you come to the edge of the pavement to go to the park, you can stand with the single lead and all the dogs yapping, but under control until the traffic relents and across the road you go. The other system is MIMD (Multiple Instruction stream, Multiple Data stream), where you have independent processors, all with their own instruction schemes (Figure 1). In this case, you have decided to take 20 dogs for a walk but you have no lead. You can just imagine the chaos that can happen.

You could try to synchronize them i.e. make them all do similar activities. But to some extent, this wastes a lot of the power of the system. Obviously the previous design (SIMD) could only do identical things. In other words, you could only all add, or only all subtract etc. using the data. Whereas in the MIMD, for instance, you could add with one processor, and subtract with another processor or use it for instruction modification. So, this is the best, and the most difficult design to aim for. Obviously, you can pipe-line processors like we can pipe-line units inside the machine (Figure 2). You can have dedicated systems where, like in task assignment processing, where with many things to do, you can have one processor to do one thing and then link it to another processor, and once a pipe-line is filled, a very fast speed is obtained. In a special dedicated system, a job or a task is split into smaller pieces which also can be overlapped.

We return now to the SIMD, array processor. Many problems are two-dimensional in nature, and these processors can communicate by a limited connection with neighbouring processors and they each have a small memory, and many partial differential equations can be solved. But the SIMD is a synchronous machine; it is not possible for one processor to do one operation and another processor to do a different operation. They must do all the same operations.

A multi-processor system can be either tightly-coupled or loosly-coupled. In other words, they can either pass messages, or they can be connected to a common memory. The difference being, for instance, if I wanted to communicate to you I could tap a wire straight into your brain and communicate with you, or I could leave a message in a post box and you could pick it up at your convenience. You have less conflict in the latter. The closely coupled ones are the designs we are currently working on to improve performance.

In one multiprocessor system we built at Loughborough, we have four machines connected to a large shared memory, each machine also has its own memory. Also, there is a disk, VDU and so on. Programing such a system bought back memories, of the early 1950s when one of the problems we had was how to build reliable memories. It was quite common to

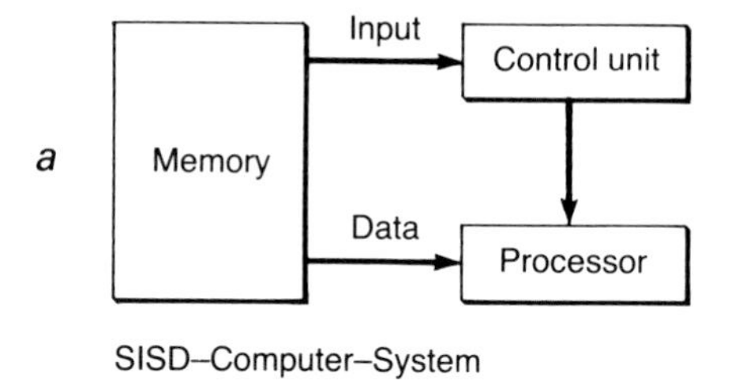

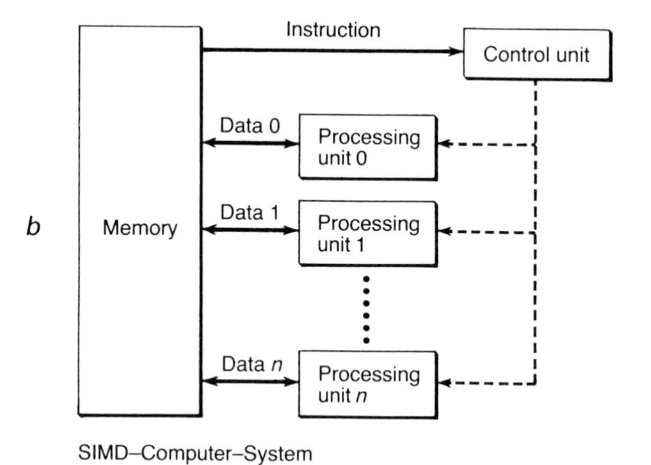

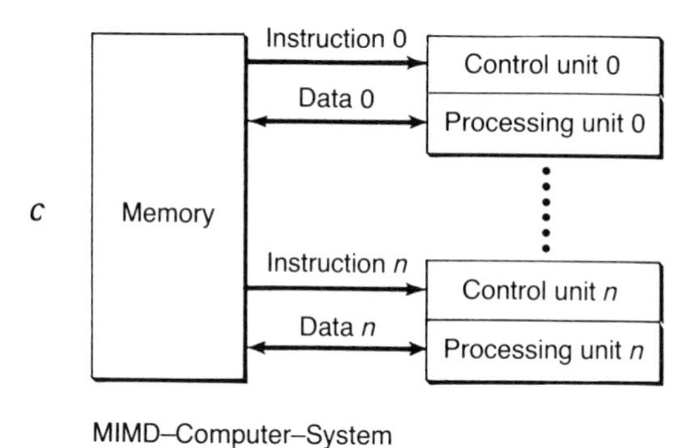

Figure 1 *a*, SISD computer system. *b*, SIMD computer system. *c*, MIMD computer system.

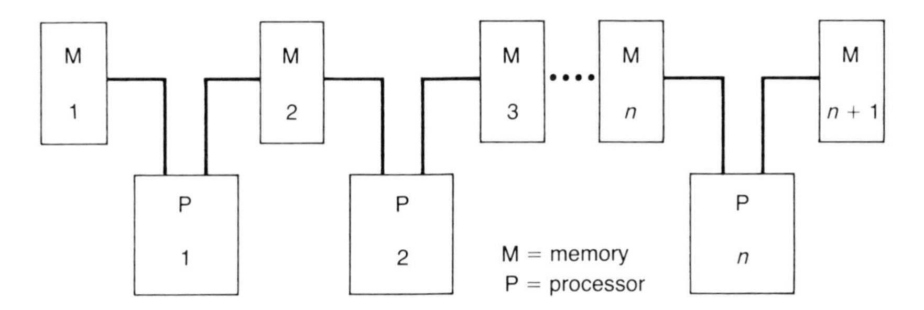

Figure 2 Pipe-line computer.

do some calculation, leave the number inside the drum, and then come back a week later and find that it had mysteriously disappeared. Then we learnt how to improve our memory design, so that it never occurs today. But with this system, for instance, one processor can actually put a number in its shared memory and carry on computing, and another one can come along, and in the same memory alter that number. In a few milliseconds later, the first machine can come back to pick up the number it left, and finds that it has been replaced by another number. And so the hazards of using systems like this are quite real, but we have learnt how to build secure systems where our computations are always correct.

In a distributed system you have a communication bus and you can put processors and memories on to produce a time-multiplexed system. These multiprocessors are influenced by the allocation and synchronization of the processes because if there is a lot of traffic on this bus, you have to wait for a response causing delays in the system. In fact, we have a system where if you have many processors and each processor has got equal priority. Then if you have one processor you have 100 per cent efficiency. But if you have two or more processors you can work out what the probability of conflicts are and when you start using five to 10 processors then it is almost impossible to avoid conflicts. This is what happens inside the machine. Figure 3 shows the ideal line, in fact, up to around about between four and eight CPUs the curve flattens off and you literally get no extra performance, and in some algorithms it actually comes down for most processors, so there is no gain. Researchers in this area are trying to push this curve up towards the maximum line. We now have better algorithms to use on these systems. In fact, we can set these machines up on a hierarchial bus structure. We have also designed them with a cross-bar switch. In other words, we do not queue on a bus, and use, say, the public highway. We have a separate line to each processor. However, its an expensive process, and current research is into what is the best network for connecting all the memories of the processors with all the input–outputs. We do not want to

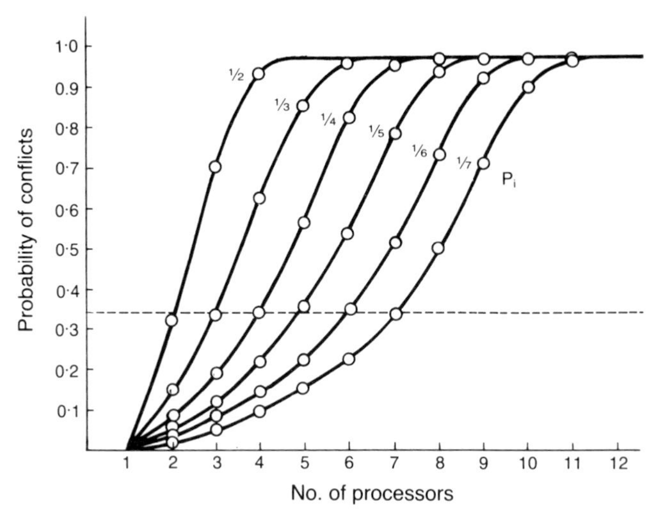

Figure 3 The probability of a delayed access in a distributed system.

connect everything to everything else, although that would be ideal, because there would be no delays—but it is expensive.

When running algorithms on these systems, we need some common bench mark. How do you test a parallel algorithm for its effectiveness? How much faster is it than the sequential algorithm? The basic definition is that the Speed Up (S_P) of a parallel algorithm using P processors, over the corresponding sequential algorithm is defined by $S_P = T_1/T_P \geqslant 1$ and its efficiency by $E_P = S_P/P \leqslant 1$, where T_1 is the time taken on one processor and T_P is the time for P processors. Then by working out the speed-up, and the efficiency, we can find out how effective our parallel algorithm is. There are also other bench marks i.e. P times T_P is the measure of the cost of an algorithm. All this leads to the topic known as algorithmic complexity.

Suppose we now want to design a parallel algorithm. Let us take a simple example of going shopping in a supermarket. Of course, a sequential task would be, you would leave home, travel to the supermarket, have a list of key items, come through the checkout, and then return home. Now then, one of the ways in which we design a parallel algorithm is the divide and conquor paradigm. I leave home and now introduce an overhead by going round and picking up some of my friends i.e. L friends. Then I divide the shopping list by L, so we all go to the supermarket and then go home. By introducing this overhead, I hope that I will gain eventually because obviously I have to pick my friends up and take them home. Hopefully, I will gain by each of them doing their own shopping sub-list. Now just imagine what can happen inside the supermarket. Let us suppose that I have divided my shopping list at random. Then one of my friends would go to one counter to get butter, another one would go to the same counter to get cheese, and there would be internal clashes. Whereas if I looked at my shopping list and divided it up into those items from the butter counter and those from the hardware counter then each of my friends would go to each separate counter and we would all gain. This is what happens in the efficient design of a parallel algorithm. We want to avoid internal clashes. They occur in the memory and also on the bus.

Let us now consider some fairly simple calculations. In the sum, AB+CD + XY then obviously all these are independent calculations because the plus sign is an associative operator, and so I can have one processor to multiply AB another one to multiply CD etc. In fact, we have parallel constructs in all the computer languages, FORTRAN, ALGOL, ADA and Pascal and we can put instructions like 'do together' instead of loops. In fact, the primary feature that distinguishes parallel algorithms and systems from the more usual uniprocessor situation, is that parallelism entails the use of facilities or resources not present in the sequential system. In multiple-processors, we have a data communication problem, and, of course, if you do synchronize the processors you waste a lot of power.

To take another approach, if we are going to look at all our programs with a view to designing parallel algorithms, we could consider converting a serial algorithm, as the first step. Well this would only give a linear improvement because most parallel algorithms will give a log N improvement. If there is no log N result then the parallel analysis must be reconsidered. Sometimes you can get a log log N improvement. There are also several principles involved in Parallel Algorithm Design. The first, is *vectorization*, another is *divide and conquor*, or *partitioning*. Another is *recursive doubling* and another is to convert *implicit forms* to *explicit forms*, and finally you can embark on designing a new parallel algorithm entirely. The field is wide open and just emerging is a new and very interesting subject called Parallel Numerical Analysis. Let us consider vectorization. If I have a vector sum $x = a + b$ these are all independent operations i.e. $x_1 = a_1 + b_1$, $x_2 = a_2 + b_2$. We have parallel compilers that can determine whether they are all independent and so we can do them all together. It is a straightforward, and an obvious step that vector computers do easily. A slightly more complex principle is matrix vector multiply. $\mathbf{Y} = AX$, and if you remember to obtain Y_1 you have to evaluate $a_{11}x_1 + a_{12}x_1$ etc. for all the rows (see Figure 4). What you do is split the number of components, so that one does P components in one processor and then another P components in another processor, etc. In fact, they can all be done at the same time. Finally, we no longer add numbers together serially, but do the parallel approach to adding numbers. If we have $a_1+a_2+a_3+a_4$ and so on, then we

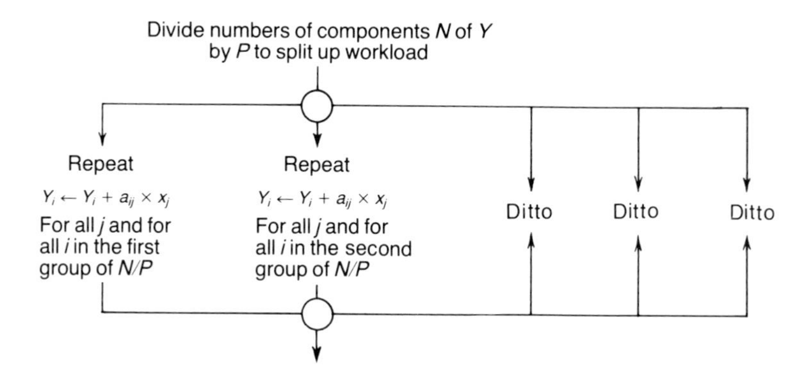

Figure 4 Partitioning of a task—Matrix vector multiplication.

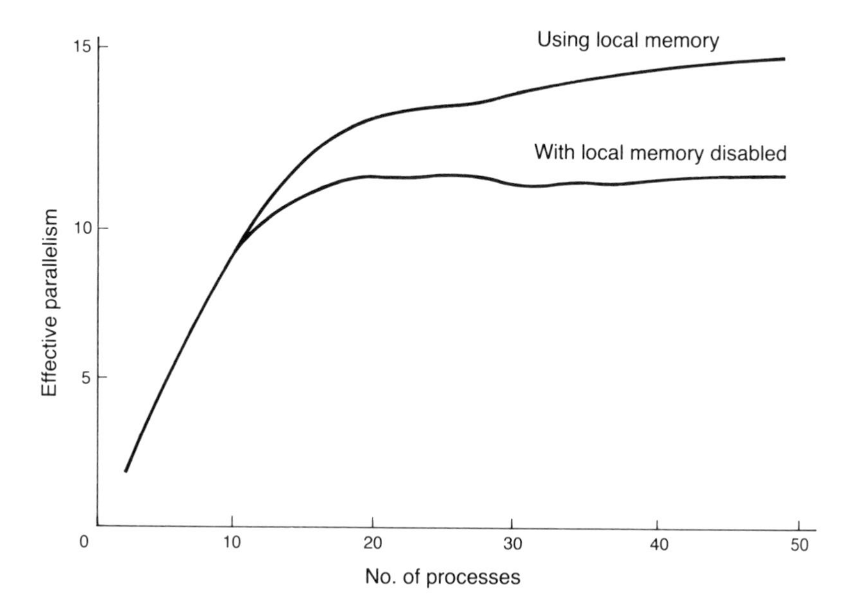

Figure 5 Number of effective processors in a parallel MIMD system.

add a_1+a_2, and another a_3+a_4 etc., This gives what we call the fan-in algorithm (see Figure 9*b*) or the log *N* improvement. Therefore, not only is the work split up into smaller jobs but you change the nature of the way we do jobs.

Take an example like quadrature. We can consider integration by following the simple paradigm of Simpson's or the trapezoidal rule. We split the integral up into a number of smaller integrals. So I have partitioned the range up into a to x_1, x_1 to x_2, x_2 to x_3, x_3 to b and provided, of course, there is no singularity or awkward discontinuity in the integral we apply the trapezoidal rule at all the points at the same time and then do a log sum addition to get the answer in two parallel computational steps.

$$A = \int_a^b f(x)\mathrm{d}x$$

Partition integral

$$= \int_a^{x_1} f(x)\mathrm{d}x + \int_{x_1}^{x_2} f(x)\mathrm{d}x + \int_{x_2}^{x_3} f(x)\mathrm{d}x + \int_{x_3}^{b} f(x)\mathrm{d}x$$

$$A_1 + A_2 + A_3 + A_4$$

$$A_{12} \qquad A_{34}$$

$$A_{1234}$$

The results on the parallel machine show a curve like Figure 5 is obtained. After 10 processors the speed up line starts levelling off, but, the curve has started flattening at between five and eight processors. It improves up to 15 and 20 and hopefully, it should continue. Some recent experiments coming from the US have gone up to about 40 or 50 processors, so 50 CPU power can be obtained but in the sequential algorithm approach we get only as little as 12 CPU power.

We can also consider a non-numerical application. Imagine you wanted to sort some numbers given in a sequential list. The algorithm chosen is the simplest of all, the bubble sort, or you could choose shell sort, or quick sort. The procedure is to search for the largest item and then interchange, put the item at the top, and continue with the process. If the list is very large the process will take a long time. But if you now split the job up between *P* processors, we then have *P* shorter sublists to search and sort, but because the list has been split up I now have to merge to obtain the final shorter list. Thus an analysis of the work is required because of this extra stage. If the total time of the job is going to be bigger than what was before then it is inefficient. Hopefully we can sort the sublists quicker and merge, so that although there is an extra overhead we still gain. This is similar to the shopping market analogy where I introduced the overhead of going to pick up friends and then having to take them home.

We have heard a lot about the Fibonnacce sequence in the conference, let us see how we would calculate this series in parallel. The series comes under the name of a second-order-recurrence relationship and a recurrence relationship is a well known sequential process in mathematics. You are given F_0 is 1, F_1 is 1, F_i is $F_{i-1} + F_{i-2}$. The way in which mathematics was taught in school is purely sequential in the sense that before you can calculate F_2 you need to know F_1 and F_0, which are given. There is no obvious way in which it can be done in parallel. However, if we now reformulate it as a first-order recurrence process and take a two by two sub-vector, f, a two by two matrix A_i, and then find the relationship $f_i = A_i f_{i-1}$. Now, whereas before in the sequential form, I would calculate F_2 and then F_3, F_4 etc. which is purely sequential. What I now do is take the global view, and say, $f_i = A_i f_{i-1}$, and then I carry on $f_{i-1} = A_{i-1} f_{i-2}$ and so on, until I reach $f_2 = A_1 F_1$. I now take the large product $f_i = A_i A_{i-1} \ldots A_2 A_1 f_1$ and do it in a fan-in calculation as before. I can do it in log N steps and then multiply by the F_1. And so you get a log N improvement with almost every algorithm.

Recurrence relation $F_0 = 1$, $F_1 = 1$, $F_i = F_{i-1} + F_{i-2}$, $i \geqslant 2$

Reformulate into a first-order process

$$f_i = \begin{bmatrix} F_i \\ F_{i-1} \end{bmatrix}, A = \begin{bmatrix} 1 & 1 \\ 1 & 0 \end{bmatrix}, i = 1, 2 \ldots N$$

$$f_i = A_2 f_{i-1} \quad = \quad (A_i A_{i-1} A_{i-2} \quad \ldots\ldots \quad A_2 A_1) F_1$$

Fan-in process

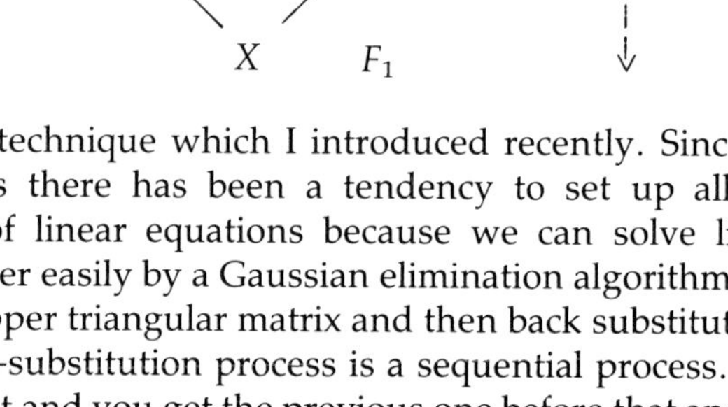

```
(A_i A_{i-1} A_{i-2}   .....   A_2 A_1)F_1
  \/  \/                        \ /        ^
  X    X                         X         |
   \  /        \               /           |
    X              X                  log N steps
      \          /                         |
          X         F_1                    v
```

There is also another technique which I introduced recently. Since we started using computers there has been a tendency to set up all our problems in the form of linear equations because we can solve linear equations on the computer easily by a Gaussian elimination algorithm. We then reduce this to an upper triangular matrix and then back substitute for the answer. Now a back-substitution process is a sequential process. You start with the last element and you get the previous one before that and the one before that etc. You now wonder, how can I parallelize that? Very simply. You just take away the implicitness of that calculation and convert it back to its original explicit form. In fact, the simple Cramer's rule was that if you worked out the discriminant of the coefficients you can actually calculate X_1 and X_2 at the same time.

$$a_{11}x_1 + a_{12}x_2 = b_1$$
$$a_{21}x_1 + a_{22}x_2 = b_2$$

Serial algorithm: solve by Gaussian elimination or LV factorization followed by back substitution process.

Parallel algorithm: calculate discriminant $\Delta = a_{11}a_{22} - a_{12}a_{21}$. Then evaluate $x_1 + x_2$ in parallel:

$$x_1 = (a_{22}b_1 - a_{12}b_2)/\Delta,\ x_2 = (a_{11}b_2 - a_{21}b_1)/\Delta$$

So you can introduce parallelism in this way also. But why did we introduce implicitness into our calculations? It has been done in partial and ordinary differential equations, for the reason that it brings more stability to our calculations. Now we are trying to take out this implicitness and make them all explicit again because once they are in explicit form you can have all these processors working on the problem at the same time. So there is a tremendous amount of work still there to be done.

Until now most of you probably thought of an algorithm as just an

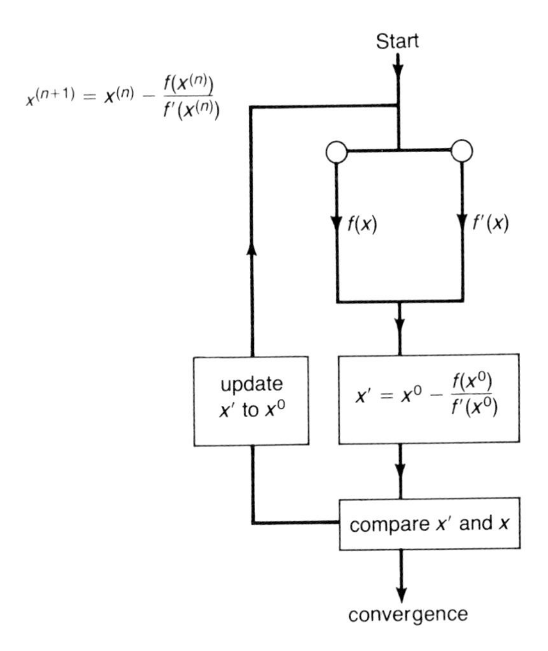

Figure 6 Newton's method—synchronous algorithm.

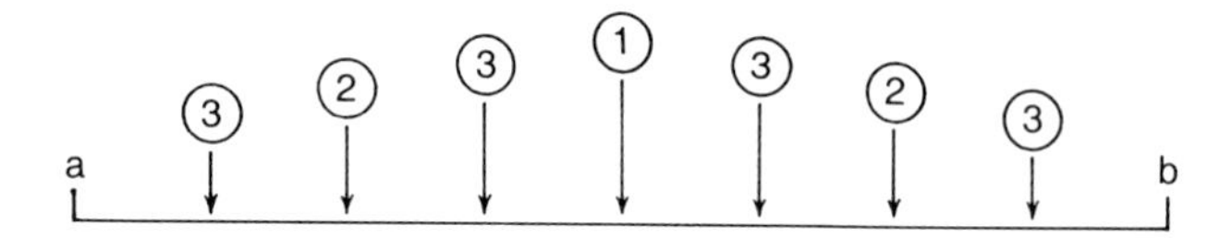

Figure 7 The bisection algorithm.

algorithm. But we have now started talking about two sorts of algorithms. One is a synchronous algorithm and the other is an asynchronous algorithm. An example of a synchronous algorithm is the very simple Newton–Ralphson Iteration, to find the root of $f(x)$. As can be seen in Figure 6 there is an $f(x)$ function evaluation and $f'(x)$ evaluation and on a parallel computer you can calculate these two at the same time. By doing this you can get a factor of two improvement almost immediately. Now the difficulty is that you have assumed that it takes the same time to calculate $f(x)$ as it does to calculate $f'(x)$ and this is not always the case. Sometimes one of these processes will finish before the other. So we have to go through a process of synchronization before we can compare and continue the iteration. Another sort of calculation which is far better for working on parallel machines is the asynchronous algorithm (see Figure 7). Again we are trying to calculate the roots of an equation. What we do is evaluate the function at the centre point of the interval and then determine where the root is in each half and then calculate at the mid-points of the sub-intervals, etc., and then at the mid-points again. All these calculations are completely independent and they can be done at the same time. And the processors can work at each one of those points evaluating the function and determining where the roots are. As we research into how best to develop new parallel algorithms there is no doubt that asynchronous algorithms have most advantages because all the processors are working at their optimum speed. Unfortunately there are very few of these algorithms, most are synchronous and, in fact, throughout the development of mathematics since the seventeenth century we have improved our algorithms by bringing in synchronization elements. I can now see us going back to rediscover and improve old methods. For instance, in the type of process known as *binary bisection*, if we had parallel processors when this algorithm was invented, it would now be known as *multisection*. This is one of the new algorithms we are investigating for two- and three-dimensional searches in Mathematical Optimization.

So a whole new philosophy is developing for constituting parallel algorithms. The characteristic of an asynchronous algorithm is that the decision tree of the process (Figure 8) always subdivides and never loops back. So you are always creating a large number of tasks for all the

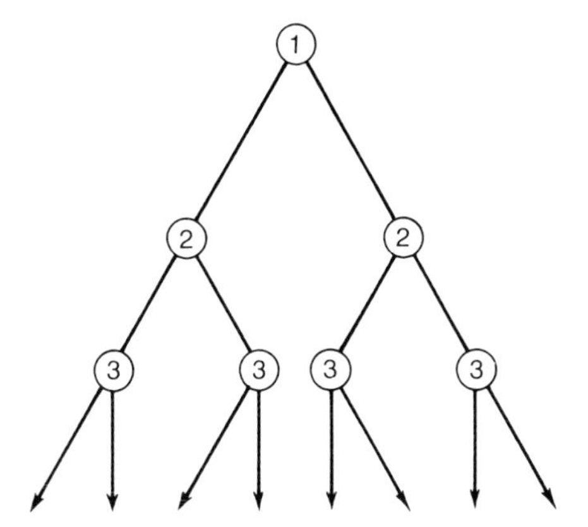

Figure 8 The parallel bisection algorithm—asynchronous algorithm.

processors to do. Consider the supermarket manager, with 12 checkouts. He does not stand at the door and only let 12 people into the shop. He lets all 100 go in, and they all queue at the checkouts, and during that time create an enormous number of tasks. In the same way, you can queue tasks through all the processors. This is how we build and design new algorithms. In other words, the problem is divided up into a large number of independent tasks all of which can then be accomplished in parallel. We do not actually design an algorithm to use two or four processors. We design an algorithm to use processors where P is a parameter in the system. This is known as the degree of *granularity*. In other words, you can divide your job up into a small number of processers or you continue until you get very tiny jobs. Very much like the convict who instead of breaking a stone up into 10 pieces, continues until he has a pile of sand. When you obtain the answers to all the many tasks you have to bring them all together which increases the communication inside the system. You must try to reduce this and the accesses to the common memory of your machine. Otherwise the overheads will make the algorithm inefficient. Figure 9*a* shows the sequential approach that we used to do for addition. We add the first two numbers and then the third and fourth, this produces a highly structured tree. Such trees are unsuitable for the parallel machines because only one processor can access it. Figure 9*b* shows the parallel approach where all the processors are working. We obtain a log N/N improvement for N terms. In fact you can also multiply numbers in the same manner in log N steps, normally it takes $N-1$ cycles. If you have matrix vector multiply you can do it in log N steps with N processors. It is normally an N^2 operation. Matrix inversion is an N^3 operation which can be reduced to N steps in parallel.

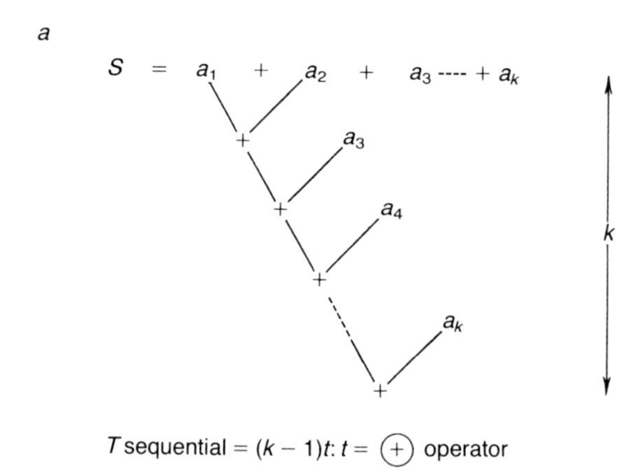

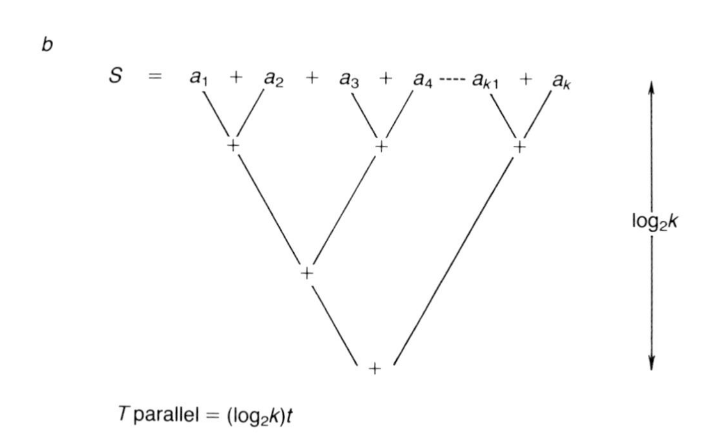

Figure 9 Addition of k numbers: *a*, sequential; *b*, parallel.

This is a tremendous advantage to problem solving because the factor N^3 virtually kills off the desire to do large jobs on our present-day machines. So we are starting again with parallel machines with one or two orders of magnitude gain. In fact, on most ordinary differential equations you can get 50 to 1 gain. So you can find new parallel algorithms for most problems, which is a most exciting challenge.

Dr. Geoffrey Kaye

Geoff Kaye is head of the research group working on problems associated with speech synthesis at the IBM UK Scientific Centre.

Born a Yorkshireman, he obtained a PhD in Nuclear Physics at Liverpool University. After a postdoctoral fellowship at Atomic Energy of Canada, Chalk River, he returned to Liverpool as a Lecturer. Eventually the lure of industry became too great, and he joined IBM at the Hursley Development Laboratory in 1968. Here he worked on a wide range of products until, in 1980, he joined the UK Scientific Centre, managing the Image Research Group. Three years ago he took up his present post.

His hobbies are sailing, Indian Postal History, antique cartography, and a love of books ancient and modern.

Orthography maid plane

All languages are spoken, but not all are written. Most native English speakers would have little difficulty with the title of this talk if it were presented to them aurally. However, on reading it, the reaction of many would be that it is nonsense. When it is correctly spelt most problems are removed, but someone may be uncertain about the exact meaning of the first word. If a literal etymology is used then it could be said to mean *right writing*. The title then becomes *Right writing made plain* which unfortunately is ambiguous when spoken. Is it a command, or the title of a paper that will elucidate spelling? In normal conversation, the intonation used by the speaker, and the larger context, would resolve the ambiguity. These are examples of something common to many languages: the potential ambiguity that is present when homophones are spoken and which is resolved when they are written. Hence, it could be said that the lie is given to those who would maintain that there should be a unique spelling for each sound in a language, and in particular for words that sound the same. That two words are now homophones does not mean that they have always been so, nor that their different spellings are illogical. Originally most homophones had distinct pronunciations, but due to pronunciation mutations, especially of the vowels, words that were originally spelt and pronounced differently have now come together as homophones. Similarly, spellings that had one pronunciation now have two or more. In Shakespeare's time *sea* and *see* did not rhyme, neither did *heap* and *keep*. Likewise, there are words in Middle English with the same long vowel that have now developed into three distinct pronunciation groups whilst maintaining the same spelling for the vowel. For example, *room, food* and *roof* by Shakespeare's time had a pronunciation close to their modern form which is distinct from that of *blood* and *flood*, but it was not until 1700 that *good, stood, book* and *foot* acquired their present pronunciation. Much of our vocabulary had acquired a spelling close to today's form by the fifteenth century, but mutations in pronunciation have continued and are still doing so, with the inevitable divergence of the spelling and the pronunciation. This process of mutation has occured, and always will occur, as long as English is a living language.

English orthography has been influenced by two maor orthographic conventions; Old English (pre-Conquest), and Norman French, with significant later contributions from Latin during the Renaissance. At the time of the Conquest there was considerable uniformity in the spelling of Old English as written in the monastic scriptoria. It was the vernacular, and the language of Church and Court. The Conquest brought a new ruling class who spoke Norman French. This became the language of the Court and the Law; and at the same time, the Church started to use Latin more in its liturgy. Consequently, the demand for books in Old English declined, and with it the spelling standards established by the scriveners were lost through lack of use. When England's contacts with France began to decline, English once more became the only vernacular, and reestablished itself as the language of the Court and commerce. Unfortunately, the intervening 300 years had seen such a large drop in the use of written English that few of the old orthographic standards remained, and those that did remain were now heavily intermixed with those of Norman French. Such influences are never uniform in their effect, and inevitably lead to inconsistencies. These inconsistencies are most marked in words that have come to us from Old English, when compared with words borrowed at a later date.

Our unashamed borrowing of words from other languages has given a richness to our vocabulary, but not without orthographic consequences. Not only do we borrow words, but quite often a word is borrowed twice after it has travelled from its source via different routes and has been subject to different sound changes on the way. Examples of such words are *brotherly* and *fraternal*, and *warranty* and *guarantee*. Borrowings have also taken place at different times from the same language, which has results in the same spelling having a different pronunciation at each borrowing. For example, words such as *cheese* and *chin* acquired the spelling in <*ch*> in about 1200 as a direct influence of Norman French, whilst borrowings from Modern French give us the pronunciation found in *chauffer* and *machine*, and Latin borrowings have given rise to *archive* and *chorus*. So, whilst the spelling-pronunciation relationship in words in each of these categories is consistent and rational, the pronunciation of <*ch*> when viewed across the whole lexicon of English appears inconsistent. Words that are initially borrowed in a literary context usually retain their original spelling, but those borrowed orally often enter the written language with a spelling closer to the English orthographic conventions at the time of borrowing.

Despite the manifest inconsistencies in English spelling we need to answer the questions, "*whom does an orthography serve, and why do we have one?*". We write to record ideas, thoughts, happenings, instructions, so that they may be communicated to others remote in time and space. This mode of communication was developed long before the advent of modern communications technology. It is interesting to speculate whether, if the reverse had been the case, writing would have become such a universal tool of civilisation. However, in a time of almost universal literacy amongst those for whom English is their first language, we must not forget that the spoken form of language predates any written form, and that writing originated as a means of recording what was spoken or thought. That this is often forgotten is manifest from the frequency with which one hears the question, *"How do you spell that?"* compared with the virtual absence of *"How do you say that?"*. Writing was developed by speakers of a language for their own use, and it may be reasonably assumed that intended readers were expected to be fully conversant with the language. Such readers can easily produce the current spoken form of what is written by applying the rules of pronunciation and syntax used when they are speaking, and also from their interpretation of the communicative intent of what is being read. With this in mind it might not be unreasonable to claim, as Chomsky and Halle[1] do, that *"English orthography, despite its often cited inconsistencies, comes remarkably close to being an optimal orthographic system for English"*. This is seen to be so when one remembers that since the intended users are fluent in the language, the orthography does not burden them with unnecessary detail of phonetic variation, when this can be provided by general rules known to the readers and which they use when speaking. Consequently, the following information is not supplied where it can be predicted by the reader. Although there are exceptions to the general statements made below, nevertheless, they are widely applicable.

Vowel reduction

Polysyllabic words usually have one syllable that is stressed more than the others. The first syllable in the first word of each of the following pairs is not stressed. Thus the vowel in these syllables has a reduced form (indistinct quality) when spoken.

*a*bandon *a*bolition
b*e*gin b*e*nefit

The first syllable in the second word is stressed, and so its vowel is not reduced.

Allophonic variations

The sounds of a language are influenced by the context in which they are spoken. Examples of these *allophonic* variations for the English sound represented by /t/ are discussed later.

Vowel alternation

There are many word pairs, such as those that follow, where the spelling does not reflect the sound change of the vowel (in bold italics) when they are spoken.

div***i***ne	div***i***nity
ser***e***ne	ser***e***nity
prof***a***ne	prof***a***nity

Such alternations are predictable and so need not be reflected in the spelling.

Consonant alternation

Some word pairs show an alternation in a consonant, rather than a vowel. The following table contains three different cases of the marking of consonant alternations.

Unmarked

confiden***t***	confiden***t***ial
expre***ss***	expre***ss***ion

Marked

permi***t***	permi***ss***ive
persua***d***e	persua***s***ion

Marked but not pronounced

licen***s***e	licen***c***e
practi***s***e	practi***c***e

In addition to these, neither *lexical stress, intonation* nor *assimilation* are marked orthographically. Examples of these are given below.

If, however, the purpose of an orthography is to enable a person not conversant with the language to read a passage such that it is acceptable to a native speaker, then we must look for a representation that is significantly more detailed than most orthographies. A suitable representation could be a detailed phonetic transcription. The two English sentences:

The electrician is adapting an electric typewriter.
It will run on the new electricity supply.

which are slightly contrived to highlight some of the items discussed above are shown below with three increasingly detailed levels of phonetic transcription.[2]

Phonemic transcription

/ðɪ ˌɛlək'trɪʃn̩ ɪz ə'dæptɪŋ ən ɪ'lɛktrɪk 'taɪpraɪtə/
/ɪt wil rʌn ɒn ðə njuː ɪˌlɛk'trɪsɪtɪ sə'plaɪ/

Allophonic transcription

[ði' ˌɛlək?°'tr̥ɪʃn̩ ɪz ə'dæp?°tĩŋ ə̃n ɪ'l̥ɛk?°trɪk?° 'tʰaɪpraɪtə]
[ɪt? wɪɫ rʌ̃n ɒ̃n̪ ðə nju: ɪˌlɛk?°'trɪsɪti' sə'pl̥aɪ]

Allophonic transcription with intonation

[ði' •ɛlək?°ˇtr̥ɪʃn̩| ɪz ə↗dæp?°tĩŋ ə̃n ɪ↗l̥ɛk?°trɪk?°↘tʰaɪpraɪtə||]
[ɪt? wɪɫ ˌrʌ̃n ɒ̃n̪ ðə ˌnju: ɪlɛk?°↘trɪsɪti' sə•pl̥aɪ||]

The speaker model chosen for this example is British English Received Pronunciation. In the phonemic transcription only the essential contrastive sounds of the language are indicated. This system requires about 20 vowels and 22 consonants. Raised bars are placed before syllables carrying primary lexical stress, and lowered bars before syllables with secondary stress. Although such a transcription uniquely identifies each sound it does not indicate how the sounds are actually realized in the context in which they are spoken. Variations in realization are marked by diacritics in the allophonic transcription, which gives the major aspects of what is to be said, but not how it is to be said. The third level of transcription shows a broad outline of the intonation that would be used when reading these sentences unemotionally, and so indicates how they are to be said.

The unsuitability of the most detailed transcription for normal written communication need hardly be emphasized. Some would maintain that the phonemic transcription could be the basis of a suitable orthography, even though it would result in homophones also becoming homographs. One such spelling scheme is the Initial Teaching Alphabet (i.t.a.) devised by Isaac Pitman. A potential problem with such an approach to English spelling becomes apparent with words such as the following:

electric	/ɪ'lɛktrɪk/
electrician	/ˌɛlək'trɪʃn̩/
electricity	/ɪˌlɛk'trɪsɪtɪ/

The orthography makes clear the relationship between the words, so enabling a reader to infer the meaning of the derived forms. This relationship is less obvious in the phonemic transcription, because of the changes in vowel quality (governed by the position of the lexical stress) and the consonant alternations.

Machine speech

Despite the plea made for the optimal nature of English orthography when used by a native speaker, one is confronted with a significant problem when trying to develop a machine to read English text (stored on a computer) and produce speech that is as acceptable, as intelligible, and as non-fatiguing as that read by a competent English speaker. For this to be done with the same accuracy as a human reader, the machine will need a built-in rule system comparable with that used by the human speaker when reading. This is the challenge of research into machine produced speech, namely to define such a system of rules. Our work to date has developed rule systems for pronunciation, lexical stress, and the grammatical analysis required as a precursor to the assignment of plausible intonation. The means of carrying out a sufficiently detailed semantic analysis to ensure that the intonation pattern and other prosodic features properly convey the full communicative intent of what is spoken are a significant distance from realisation. Some of the solutions to these problems, and the considerations that must be taken into account if high-quality synthetic speech is to be produced, will now be examined.

In speech synthesis, we are interested in the relationship between two disparate, but equally complex systems: the orthographic representation of English and the acoustic signal of speech; and also in the question of how for English the first can be converted to the second. There are two other related systems with which we shall not be concerned: namely the interpretation of the orthography when reading silently, and the interpretation of the acoustic signal by the ear and the brain.

Figure 1 shows a simple English sentence and the acoustic signal produced when it was spoken by the author. We now have two representations of the message, the acoustic and the orthographic. Are they equivalent, and are they equally complex? One is tempted to say *yes* since they both carry the same message. But do they? There is no doubt that the

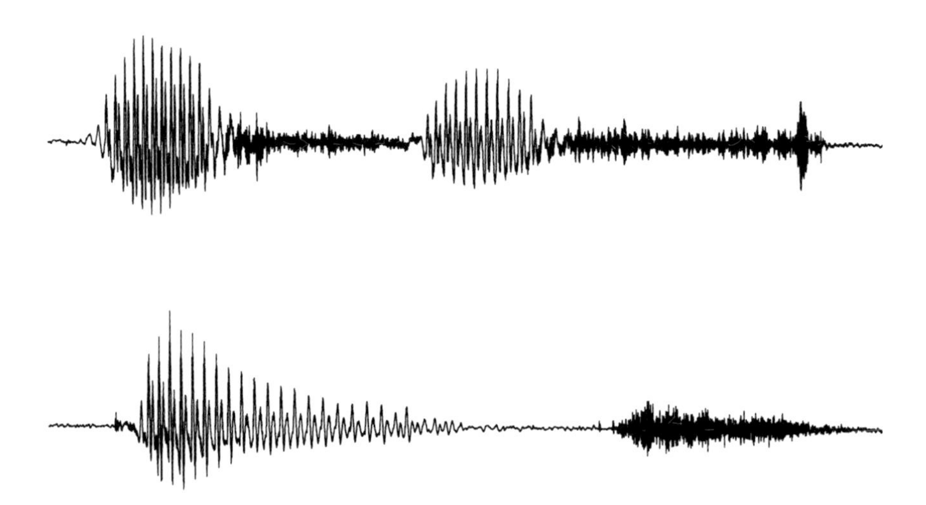

Figure 1 The sentence "This is speech".

words are the same, but the way in which they are spoken will very much depend on the reader's view of the communicative intent of the message. From the viewpoint of data transmission the text is significantly simpler. It is assembled from about 60 symbols (upper- and lower-case letters and punctuation). This text string can be transmitted as 124 bits, whereas the speech waveform will be closer to 250 000 bits. On this basis one would conclude that either the speech signal is highly redundant or that the two are not equivalent. The reality is somewhere between these extremes. If we are only interested in receiving the speech message and are not interested in fidelity then very significant reductions can be made in the number of bits in the signal (note that we are only discussing the reduction of bandwidth and the accuracy of digitization). On the other hand, if we are interested in a level of fidelity comparable to a digital recording, the amount of data will increase by another order of magnitude. Is all this acoustic signal required by the hearer? This problem is not fully understood, but there are indications that not all of the signal is used by every listener, and that different listeners may use different parts of the signal.

Let us now look at the orthographic signal. This is apparently very simple, but for it to be satisfactorily spoken the reader has to apply a large set of rules to arrive at the correct pronunciation at the phone (sound) level. In addition, the communicative intent of the message has to be understood from the intrinsic content of each sentence, and at times its larger context, to enable the reader to use the appropriate intonation pattern. It is therefore necessary to apply two rule systems: one to determine what to say, the other to determine how to say it. It is tempting, and plausible, to postulate that the orthography, together with the interpretive rules, is a system of equal complexity to the acoustic signal.

What has to be done if we are to produce a reading machine that will read text in conventional English orthography (stored on a computer), and produce speech that is intelligible, acceptable and non-fatiguing? We have to discover and codify the rules that we, as native English speakers, use when we read an English text aloud. This has proved to be far from simple. Unlike the laws of the physical world, we are dealing with a rule system that is an artefact, and that consequently exhibits the vagaries and idiosyncracies seen in any organism. The sceptic might ask, *"are there any rules relating English spelling and its pronunciation?"* We all know some spelling rules: for example the rule defining the sequence of *e* and *i* in the context of *c*. If the sound is like *e* put *i* before *e* except after *c*. Many such rules exist, and we will look at a few of the less obvious. Before doing so, it is important to appreciate that it is not essential to have a one-to-one correspondence between each sound in a language and a symbol in its alphabet. If this were to be so, then English and many other languages would require significantly larger alphabets, with English needing about 42 characters depending on the accent. Sequences of letters, (not necessarily consecutive) can be used to represent a sound, and more than one sequence may be used for the same sound without ambiguity. For example *meet, meat* and *mete*. Unfortunately, there are many instances where the same letter sequence is used for different sounds, leading to ambiguity in pronunciation. Examples are: *head, steak* and *meat*, and the notorious *-ough*. A satisfactory rule system for English must identify rules of general applicability, and also list the exceptions to the rules so that they can be handled separately. The following two orthographic rules have general applicability but are less obvious than the one above.

The vowel *a* in *bag* is pronounced /ɒ/ as in *bog* when it is preceded by /w/ as in *wit* and followed by any consonant other than /k/, /g/ or /ŋ/ which are respectively the terminal consonants in *thick, dig* and *ring*. Typical examples are:

wad	/wɒd/
wag	/wæg/
quack	/kwæk/
quantum	/kwɒntəm/
what	/wɒt/

This rule is however subservient to the so-called *magic-e* rule which results in pronunciations such as:

wade	/weɪd/
wage	/weɪdʒ/
quake	/kweɪk/
persuade	/pəsweɪd/

The operative rule is: when a word ends in the sequence *vowel-consonant-e* the vowel before the consonant is pronounced with its alphabetic name, and the *e* is not separately pronounced. This is sometimes referred to as a silent *e*. It is not silent, but is pronounced as part of the preceding vowel in the above sequence. It should also be noted that there are four different orthographic representations for the sound /w/ in these two sets of examples. Much to the regret of foreign learners of English, and of those who wish to make machines read English, there are exceptions to rules such as these which must be handled as special cases.

It was mentioned above that often consonant alternations are not explicitly marked; for example,

fanati*c*	fanati*c*ism
toxi*c*	toxi*c*ity
classi*c*	classi*c*ist

Here the rule is that: /k/ becomes /s/ when the suffix begins with /ɪ/. The following pairs of words follow a similar rule

democra***t***	democra***c***y
subver***t***	subver***s***ive
permi***t***	permi***ss***ive

where /t/ becomes /s/ when the suffix begins with /ɪ/.

Similar rules apply to other cases of consonant alternation governed by suffixation. None of these alternations need be shown in the orthography since they are all predictable, and known (albeit subconsciously) to all competent speakers of English. Some pronunciation rules require knowledge of the syntactic context in which the word occurs. Words such as *house, abuse* and *refuse* have a consistent alternation in their final consonant depending on whether they are used as verbs or nouns. For example:

Where is the chicken house?
Will you house the chickens?

Other words change the stressed syllable, and in some instances the vowel in a syllable. For example:

How much is the discount?	/'dɪskaʊnt/
Do you discount this?	/dɪs'kaʊnt/
I am absent.	/'æbsənt/
I absent myself.	/əb'sɛnt/
The new record.	/'rɛkɔːd/
Record it anew.	/rɪ'kɔːd/

To pronounce these words correctly it is essential to know their part of speech in the utterance. Hence the underlying operative rule system is a subset of the grammar of English.

We are all familiar with the confusion that foreign speakers of English have with words like *record*, but have little difficulty in understanding them when they misplace the lexical stress. However, English lexical stress is far more complex than in most other European languages, and is not as easy to predict as might be assumed from the above discussion. It can give rise to waggish humour as in Figure 2, but could result in miscommunication in a sentence such as:

Mr Smith is important at work

where incorrect lexical stress could result in *important* being pronounced as *impotent*. The automatic prediction of lexical stress has been demonstrated by my colleague B.J. Williams.[6]

Figure 2 A humorous use of English lexical stress

Despite the complexity of, and the many exceptions to, the spelling/pronunciation rules of English it is possible to develop a system of rules for the automatic phonetic transcription of an English text. Notable previous work in this area by Ainsworth,[7] Elovitz,[8] and McIlroy[9] has resulted in rule systems with varying degrees of success. These approaches all follow a similar strategy of making a single left-to-right scan of the orthographic string, and produce a broad phonemic transcription for each orthographic character or character string, based upon the preceding and following orthographic context. We have recently made several advances in the development of this approach to automated phonemic transcription, as described in Lawrence *et al*;[10] in particular by allowing, where appropriate, the phonemic string so far generated to supply the left context in relation to which the orthographic string is examined when determining its phonemic transcription.

An example of a small section of the rules associated with the transcription of the letter *e* are shown in Table 1. This shows the context testing rule, and the output phonemic string generated if the rule is satisfied. Each rule consists of three parts:

(1) The orthographic string to be transcribed, shown in parentheses. It may be any number of letters, or even whole syllables or words.

(2) The left context, which can be either letter or phoneme strings, or sequences of letter categories, such as, vowels or consonants. The notation has the following meanings:

C1	one consonant
C1M	one or more consonants
V1	one vowel
[/k,g,ʃ,.../]	any one phoneme in the set
[t,d,..]	any one alphabetic letter in the set
a,b,c,...	a single lower case letter represents itself
#	word boundary

(3) The right context, which can have the same components as the left context but excluding phoneme strings, since this is a left-to-right scan.

In Table 1, the rules give the context for the transcription of word final *et, es* and *ed* sequences. There are also example words for each rule. The rules are critically ordered attempting to transcribe larger contexts first.

Table 1 *An example of some of the rules associated with the letter* e.

Rule	*Output*	*Example*
C1M(e)g,e	= ɪ	collEge
(easy)#	= iːzɪ	quEASY
[/k,g,tʃ,dʒ/](et)C0M,#	= ɪt	budgET
C1,C1(et)C0M,#	= ɛt	regrET
[/z,tʃ,ʃ,ʒ,dʒ,s/](es)#	= ɪz	matchES
[/p,t,k,f,θ/](es)#	= s	flukES
V1,C0M(es)#	= z	gamES
V1,C0M,[t,d](ed)#	= ɪd	datED
[/p,k,f,ʃ,s,θ/](ed)#	= t	dupED
V1,C0M(ed)#	= d	lovED
V1(e)#	=	toE
(e)C1,i	= iː	Evict
(e)n,C1	= ɛ	fEnce
(e)	= ɪ	bEtween

Typical rule systems that have been investigated may contain as many as 700 rules. The following sentence

The very first showroom is used by young mums with handbags

is phonemically transcribed by these rules as

/ðə vɛrɪ fɜːst ʃəʊruːm ɪz juːzd baɪ jʌŋ mʌmz wɪð hændbægz/

If this transcription is synthesized the result will be considerably worse than 'Dalek talk', and at times will be unintelligible to unaccustomed listeners. This largely results from the fact that the rules have:

(1) Assumed that a particular phoneme has only one pronunciation irrespective of its context.

(2) Not taken account of the following phonetic context due to the scan being one-pass left-to-right.

The effect of the phonemic context on the transcription is exemplified by the variations in the pronunciation of the phoneme /t/ in the following sentence and its associated transcriptions:

eight boys had tea under the tree on the eighth day

Phonemic transcription
/eɪt bɔɪz hæd tiː ʌndə ðə triː ɒn ðɪ eɪtθ deɪ/

Allophonic transcription
['eɪt°'bɔˑizz̥æ̥æˑd°'tʰiːʌʌ̃ndəðə'triːɒn̪ðiˑeit̪θ'd̪ei]

The allophonic transcription corresponds to what a speaker of English Received Pronunciation might produce at normal reading speed. The *t* in the sequence *eight boys* is essentially not sounded, or at least not released as it would be if *eight* were pronounced in isolation. This is denoted in a more detailed allophonic transcription by /t°/. In *tea*, it can be observed by holding the hand close to the mouth that the /t/ is accompanied by significant aspiration (expelling of air). This is represented by /tʰ/. On the other hand, in *tree* there is significantly less aspiration (depending on how the *r* is pronounced). This allophone of /t/ is pronounced with the tip of the tongue behind the upper gums, but in *eighth* the /t/ (not explicitly marked in the orthography) is pronounced with the tip of the tongue touching the teeth, in anticipation of the following *th* (/θ/). This is denoted in the transcription by /t̪/. These four different realizations of the phoneme /t/ are called *allophones* of /t/. Whilst the different allophones of a phonemem contribute to the *naturalness* or *nativeness* of a particular dialect of English, they do not have the contrastive properties of phonemes, as for example the /p/ in *pat* and the /b/ in *bat*.

To produce the allophonic transcription from the phonemic transcription, a second rule system examines the context of each phoneme to give the following allophonic transcription of the sentence *The very first show…*:

[ðəverɪfɜːstʔʃəʊr̫uũːmɪzjj̫uːzd˚baɪjʌʌ̃ŋmʌʌ̃mzwɪðð̥hææ̃nd˚bægzz̥]

Items to note in this transcription are the non-release of /d/ in *handbags*; the devoicing of the final /z/ in *bags* denoted by /zz̥/ (giving a pronunciation which could be represented orthographically as <*bagzs*>); and the labialization of /j̫/.

Another feature present in normal speech is assimilation, which occurs when a phoneme is deleted or assimilated into an adjacent phoneme. This is typified by *handbags* where the non-release of the /d/ results in /n/ being pronounced in the context of /b/. There is a nasal consonant /m/ made with the lips in the same position as /b/, and so the /n/ is pronounced as /m/ when in the context of a following /b/, as when *handbags* is spoken in connected speech. Similarly, *first showroom* is pronounced as if it were spelt <*firshowroom*>. The contexts in which assimilation can occur are well defined, but it is far from simple to specify when assimilation should take place. If every assimilation were to be applied the resulting speech would sound almost drunken. The amount of assimilation that takes place depends upon the speed and formality of the speech. The following transcription of the sentence *The very first showroom…*;

[ðəverɪfɜːʃʃəʊr̫uũːmɪʒ̫uːzbaɪjʌʌ̃ŋmʌʌ̃mzwɪðð̥hææ̃mbægzz̥]

shows the result of allowing the assimilation rules to operate in all posible circumstances.

This rule system is capable of producing accurate phonemic transcriptions when supported by a modest exceptions dictionary. However, the rules do not predict lexical stress, and so inadequately predict vowel reduction, which is largely governed by the position of the stressed syllable. Neither do the rules handle words such as *house* and *record*. These problems are more easily handled by a transcription strategy based upon a morphophonemic analysis using a morph lexicon containing baseform pronunciations for each morph. The pronunciation of a compound word would be deduced using phonological rules similar to those used for consonant alternations. This strategy at the same time allows the lexical stress rules developed by Williams[6] to be used to predict correct lexical stress and to perform vowel reduction.

Even when an accurate phonetic transcription has been produced, with lexical stress fully marked, the resulting speech will be far from acceptable, since no account has been taken of intonation. The most important intonational cue is variation in the pitch of the speaker's voice. Any native speaker knows instinctively how the simple response *yes* to a question can confirm, question, express surprise, or imply innuendo, depending on the intonation used. The old adage *it's not what you say, but how you say it that matters,* was never more true. It is at this stage that English orthography gives very few clues to the person reading aloud. At times, significant clues may be given by the punctuation. For example, the punctuation of the two following sentences:

> Before the king rides his horse, he gives it a groom
> Before the king rides, his horse takes ages to groom

gives clear clues to the reading of the sentence, unlike that in the sentence below, which leads to an erroneous intonation pattern.

> *Before the king rides his horse, takes ages to groom

Other cues to intonation can be obtained from the syntactic structure of the sentence; but at times the intonation can only be decided when the communicative intent of the sentence is known, either from its semantic context, or from 'stage directions' given by the author. This is typified by the following tag questions;

> 'He's a good boy, isn't he'. said John approvingly.
> 'He's a good boy, isn't he'. said Jim questioningly.

It can, therefore, be assumed that the claim of Chomsky and Halle is probably accurate when discussing the basic sounds of the language. However, to describe intonation adequately both the semantic context and the communicative intent of the text has to be examined. This area is being given significant attention in the attempt to synthesize very high-quality non-fatiguing, intelligible, human-like speech from English text.

References

1. Chomsky, N., and Halle, M., *The Sound Pattern of English*, p.49, Harper and Row, London (1968).
2. The phonetic transcriptions are given in the International Phonetic alphabet. A brief introduction to this is to be found in dictionaries such as Collins,[3] and Jones.[4] For a more detailed discussion consult Ladefoged.[5]
3. Collins, *Dictionary of the English Language*, William Collins, London (1979).
4. Jones, D., and Gimson, A.C., *English Pronouncing Dictionary*, Dent, London (1982).
5. Ladefoged, P., *A Course in Phonetics*, Harcourt Brace Jovanovich (1982).
6. Williams, B.J., *A Metrical Algorithm for Lexical Stress in English*, Acoustical Society of America (1985).
7. Ainsworth, W., "A System for Converting English Text into Speech", *IEEE Trans. Audio Electroacoustics, AU-21.* (1973).
8. Elovitz, H.S., Johnson, R.W., McHugh, A., and Shore, J.E., "Automatic translation of English text to phonetics by means of letter to sound rules", *U.S. Naval Research Laboratory Report 7948* (1976).
9. McIlroy, M.D., "Synthetic English speech by rule", *Computing Science Technical Report 14*, Bell Laboratories (1974).
10. Lawrence, S.G.C., Williams, B.J., and Kaye, G., "The automated phonetic transcription of English text", *Proc. Inst. Acoustics*, 6(4), 249 (1984).

Dr. Geoffrey Manning

Geoff Manning is Director of the Rutherford Appleton Laboratory, the largest of the Science and Research Council establishments.

At Imperial College, London he obtained a first in Physics (1952) and a PhD (1955) on the measurement of lifetimes of nuclear states in the range 10^{-12} to 10^{-14} seconds. From Assistant Lecturer in the Physics Department of Imperial College, he joined the English Electric Company in 1955 as a research worker, then went to Canada for research work using reactors in nuclear structure physics at the Canadian Atomic Energy Authority, and later went to the California Institute of Technology for further research on the magnetic moments of short-lived nuclear states.

Returning to the UK in 1960, he joined the Atomic Energy Research Establishment at Harwell, researching in high energy physics, largely on the measurements of proton proton scattering, during which time he spent two years working at CERN, Geneva.

In 1966, he joined the Science Research Council's Rutherford Laboratory as a Group Lead working both at Rutherford and CERN. In 1969, he became Head of the High Energy Physics Division, and Deputy Director of the Laboratory. He moved in 1975, to head the Atlas Computing Division, and became Project Leader for the construction of the Spallation Neutron Source in 1977.

In 1979, he was appointed Director Rutherford in the combined Rutherford and Appleton Laboratories, leading to his present appointment in 1981.

Quarks to quasars

The Rutherford Appleton Laboratory (RAL) is an establishment of 1500 staff funded by the Department of Education and Science through the Science and Engineering Research Council (SERC). It exists to provide support for university research. The scientific programme covers most of the areas supported by the SERC and spans the full distance domain covered by advanced research—from quarks to quasars, from 10^{-16} to 10^{29} cm, a range of 45 decades. I will give a brief overview of the Laboratory and then illustrate our programme by selecting three items from the very extensive portfolio that we support.

Overview of RAL

RAL is situated 14 miles south of Oxford and is within 2 hours car journey of universities in the southern part of the UK. Even universities as far afield as Edinburgh can reach the Laboratory in about 3 hours. These, apparently unimportant, factors have great significance since we provide support for more than 4000 external users spread over all UK universities and many polytechnics.

The RAL has a budget of about £55 million per year and this very large resource is deployed on a programme determined by the SERC, its boards and committees. No part of the budget is provided for unspecified programmes of the director's choice. Some 93 per cent of the manpower is used on the provision of support for university research. Only 7 per cent is used on in-house research, the vast majority is in direct collaboration with university groups and all is subject to peer review.

The support of university research is our main role and this is undertaken as a collaborative venture and not with a servant/master attitude. The SERC spends about half of its budget on providing support directly to the universities as grants or post-graduate training awards. The remaining half of the budget is spent either in provision of central support through the SERC establishments (about 30 per cent of the total SERC budget) or through 'membership fees' to international facilities (about 20 per cent of the total budget) where the research is too expensive to undertake on a national scale. RAL supports this programme in a number of ways. The first is through the development, construction and operation of central facilities—examples include a pulsed neutron source and a high-power pulsed laser facility. A second role is to provide support for the university use of the international facilities—the 'membership fees' do not provide for exploitation. A third role is that where grants given to the universities are part of a coordinated and promoted programme with the technical support often provided through RAL. In all of these tasks, our job is to collaborate with the universities and to encourage them to collaborate with each other—a task requiring a degree of experience and skill!

A secondary role, but one that we take seriously, is where it is relevant to collaborate and/or aid UK industry. We are not in the business of general contract research but, where relevant to our overall programme, we do such work with the result that about 10 per cent of our budget is from non-SERC sources.

To undertake the support of advanced research it is imperative that we also participate in that research and, as I have said, about 7 per cent of our manpower is used in that way. About one-fifth of our staff spend typically one-third of their time undertaking research. This level is probably on the low side and more could be justified but there is always serious pressure on the resources we have available.

What then is the scientific programme we support? It includes all areas covered by the SERC. We have work for all four boards; Astronomy Space and Radio, Engineering, Nuclear Physics and Science. Table 1 gives a very brief summary of the areas in which we work for each board listed in terms of the distance scale being studied. A range covering 40 decades in distance is included. It is impossible here to cover this programme in even a cursory way and hence I will illustrate it by taking three examples spanning the spectrum that we cover. These examples will be taken from high-energy

Table 1 *Brief summary of the programme of RAL for each of the four boards of the SERC*

Board	*Programme*	*Typical size* (cm)
Nuclear Physics	High-energy physics	10^{-16}
Science	Pulsed neutron source	10^{-8}
	Laser facility	
Engineering	Information technology	1
	Radio propogation	
Astronomy, Space and Radio	Astrophysics	10^{25}
	Geophysics	

physics; our laser programme and our space programme for astrophysics. Even in these areas it will be a brief review of one small part of the programme.

Support of high-energy physics

Our role in the support of high-energy physics includes: the development of new techniques; the design, construction, operation and analysis of experiments undertaken at accelerators at the CERN Laboratory in Switzerland, in the USA and at the DESY Laboratory in Germany; and work in the theoretical understanding of the subject. In all areas, the work is collaborative with universities. I will now attempt to illustrate our role by briefly describing one experiment and summarizing the current state of play in the basic quest of high-energy physics (HEP)—the understanding of the basic laws of force that operate in our universe.

The experiment from our HEP programme that has yielded the most important results is easy to select—it is the search for the intermediate vector bosons by studying proton–antiproton interactions. Their success in finding the $W^{\pm}$ and Z° has already been recognized by the award of the Nobel Prize for physics for 1984 to Carlo Rubbia and Simon Van der Meer. The experiment was undertaken at CERN by a large collaboration including Birmingham University, Queen Mary College, London and RAL. The basis of the experiment is also easy to explain; bunches of protons and antiprotons, with an energy of 270 GeV per particle, counter rotating in the 6-km circumference SPS accelerator at CERN are caused to cross in a vacuum pipe in the middle of a 2000-ton detector. This detector is constructed to be able to observe charged or neutral particles created when a proton and antiproton interact. The angular coverage is made as complete as possible, the direction, energy and type of particle are measured wherever possible. The signature for a charged intermediate vector boson ($W^{\pm}$) is that it decays immediately into an electron and a neutrino or into a muon and a neutrino. The electron and muon are detected and recognized but the neutrinos escape, since they have such a small interaction cross-section, carrying off energy and momentum. The data are recorded and then the 'events' are 'scanned' on a computer to select those in which an electron or muon are identified with a transverse momentum greater than some cut-off value, and where there is a missing transverse momentum greater than a second cut-off value. This reduced number of events is then studied in more detail and the mass of the particle which could have decayed into a charged lepton (electron or muon) and a neutrino is computed. In practice, a very clear signature is found with a particle mass of 81 GeV with an extremely low background.

The signature for the neutral intermediate boson (Z°) is even clearer. It decays into a pair of oppositely-charged electrons or muons. Each of these

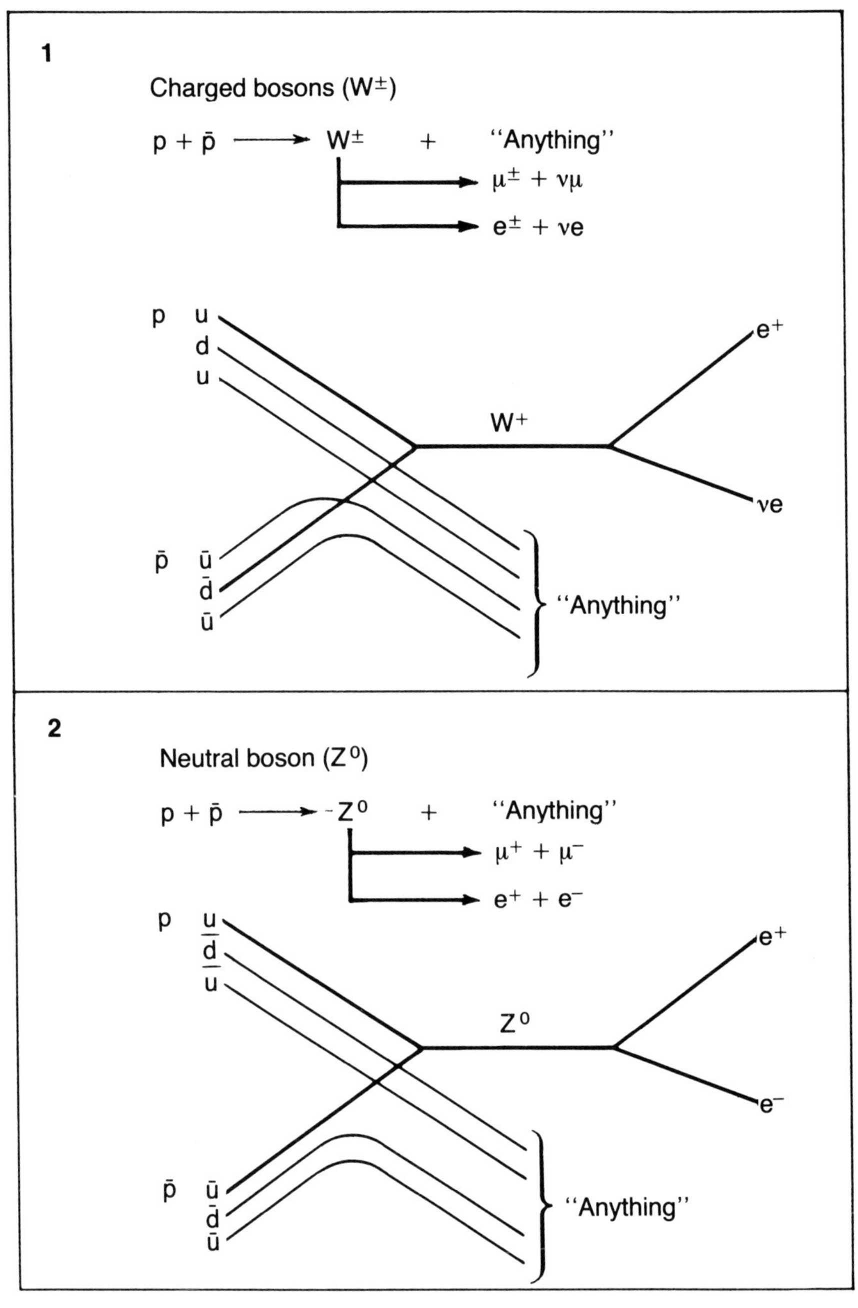

Figure 1 Diagram of reactions used to produce the intermediate vector bosons from proton–antiproton annihilation.

particles can be identified and their energy and direction measured thus allowing the mass of the parent Z° to be determined. Again a clear signature on a very low background was determined with a mass of 93 GeV.

The detailed angular distributions, the mass of the $W^{\pm}$ and that of the Z°, the relative production cross-sections, the branching ratio into electrons and muons are all compatible with the predictions of the standard electro-weak theory. There is now convincing evidence that the hereto separate theories of the electromagnetic interaction (quantum electro dynamics—QED) and the weak interaction can be described within a single unified description—the electro-weak theory.

What then is our current understanding of the basic structure of matter and the controlling forces? Table 2 summarizes our current knowledge of the building blocks of matter—the quarks and leptons. There are three quartets of leptons and quarks. The lightest group are the building blocks of the whole of the stable universe. The heavier families only come into play when unstable particles are formed. If higher excitation energies are available it is possible, but not necessary, that other heavier quartets could exist. Evidence from astrophysics places some limitation on the number of such quartets through the energy carried by diferent types of neutrino formed in the original hot big bang that created the universe, but such arguments are model dependent and are indicative rather than compelling. Figure 1 shows in pictorial form the interaction of the protons and antiprotons to form the $W^{\pm}$ and the Z°. The primary interaction is the combination of a quark–antiquark pair to form the intermediate bosons which then decay into a lepton antilepton pair. The W^{+} is formed, for example, by the combination of a u quark (charge 2/3) with an anti-d quark (charge 1/3) while the Z° is formed from a $u\bar{u}$ pair or a $d\bar{d}$ pair.

Table 2 *Summary of the leptons and quarks that are the building blocks of matter*

Lepton mass (MeV)	*Leptons*	*Quarks*	*Quark mass* (MeV) (model dependent)
$< 5 \times 10^{-5}$	ν_e	u*	~ 5
0.511	e^-	d*	~ 8
< 0.5	ν_μ	c*	~ 1500
105.66	μ^-	s*	~ 200
< 164	ν_τ*	t*	~ 40 000
1784	τ^-	b*	~ 4700

*Not seen as free particles.

Table 3 is a summary of our understanding of the forces and the mediators that carry the forces between interacting particles. As already explained the conventional four forces of yesterday have been reduced to three by the verification of the electro-weak theory—the $W^{\pm}$ and Z° have to be added to the photon as mediators. For the future there are Grand Unified Theories (GUTS) that attempt to include the strong forces with the electro-weak. These theories allow quark–lepton transitions and predict that the proton should be unstable—albeit with a lifetime in the region of 10^{32} years. There are even more ambitious plans to include the gravitational forces as well and have a supersymmetry (SUSY) that results in a single theory that explains all phenomena within the universe. There are even some brave souls that believe there is some indication of evidence to support some of the predictions of such models.

Table 3 *A summary of the forces (a) and the mediators (b) that govern the phenomena observed in the universe*

a Forces

Past theories	*Current theories* (mediators)	*Future theories* GUTS	SUSY
Gravitational	Gravitational (Graviton)		??
Weak	Electro-weak (γ, $W^{\pm}$ Z^0)	?	
Electro-magnetic			
Strong	Quantum-chromodynamics (Gluons)		

b Mediators

Mediators	*Measured mass* (MeV)
Photon (γ)	$< 3 \times 10^{-33}$
Charged bosons ($W^{\pm}$)	80 800 ± 2700
Neutral bosons (Z^0)	92 900 ± 1600
Gluons	?
Gravitons	?

RAL laser programme

The RAL laser facility was started in 1977, and it now provides facilities for about 150 users spread over many disciplines including; physics, chemistry, biology and materials science. There are four main aims:

(1) Study laser–plasma interactions
(2) Study the physics of laser-induced fusion
(3) Develop new short wave length lasers
(4) Use lasers to study a broad range of science.

Here I can address only the second item and even that in outline.

The basic reaction in fusion is
$H^2 + H^3 \rightarrow n + He^4$ with the release of 17.6 MeV

One gram of this deuterium/tritium fuel is equivalent to 10 tonnes of conventional fuel. The process is known to work, for example, in stars or in the hydrogen bomb. The only problem is how to do it on a small controlled scale. There is research on two main procedures; magnetic confinement of the hot plasmas and inertial confinement.

It is the latter that is studied in the laser programme. The basic idea is that the deuterium and tritium fuel is placed inside a small hollow shell and high-power laser beams are focused on the outside of the shell. The outer surface is ablated away and in the process drives the shell inwards by rocket action thus compressing and heating the fuel. The RAL programme is concerned with studying the physics of the process rather than demonstrating that more energy is released from 'burning' than is used in compression and heating. This aim can be achieved by using shells of 0.1–0.3mm diameter whereas at least an order of magnitude larger shells are expected to be needed to reach the so called break-even point where the energy generated equals the input energy. The laser power required scales as the cube of the initial radius of the shell and hence the laser power requirements for the RAL programme are kilojoules delivered in a nanosecond (powers of 10^{12} W) rather than many megajoules for the break-even experiments being undertaken in the USA. The lower power requirements of the RAL programme are carried out with a very flexible system that can tackle many approaches while the larger programmes are inevitable more directed and rigid. The RAL programme studies a range of basic physics in the compression process including:

What laser wavelength should be used?
What should be the shell structure?
What uniformity of illumination is required?
Are there instabilities in the compression process?
Development of computational models.

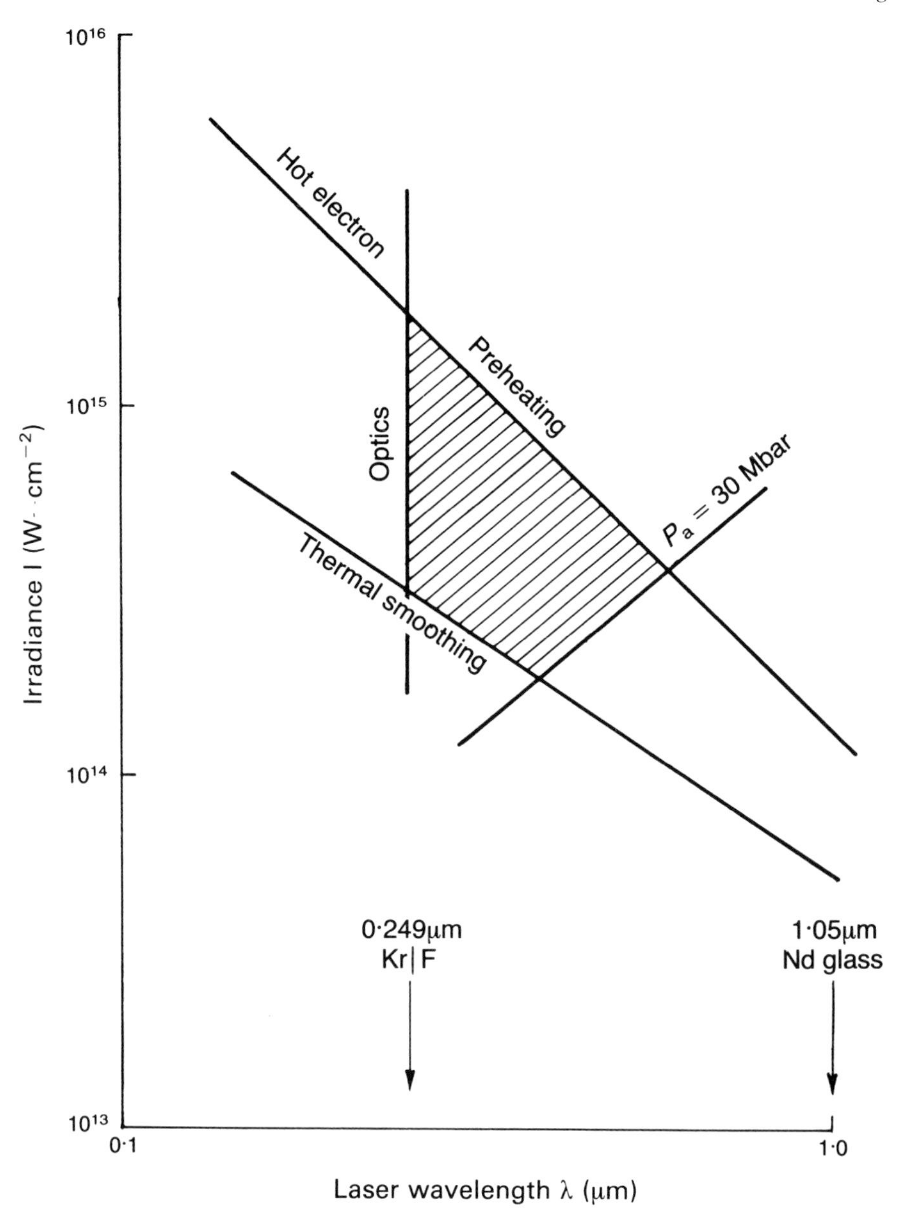

Figure 2 The shaded region shows the optimum region for attempting to get controlled fusion using laser inertial confinement. The region boundaries are drawn using experience gained from a larger number of basic experiments undertaken at the RAL laser facility.

The programme has already shown that short wavelengths (0.25 micrometres) are to be preferred, power densities of 10^{15} Wcm^{-2} and pressures greater than 30 million atmospheres are necessary. Figure 2 shows a brief and simplified summary of the results obtained. The shaded region indicates the area where inertial confinement studies seem to have the greatest chance of success and, to date, no fundamental difficulty has been found although a great deal of practical problems still remain.

Infra-red astronomical satellite (IRAS)

The third example of the RAL programme that I will cover is taken from the astrophysics area. To put the specific experiment selected in context let me first indicate some of the major questions that confront astrophysicists:

What is the origin of the universe and how has it developed?
How do galaxies form, develop and die?
How do stars form, develop and die?
Is our planetary system unique?
Are the laws of physics the same at very large distances as at very small distances?

The study of astrophysics from the Earth's surface is limited to using the radiation that penetrates the atmosphere of the Earth. This is largely visible light which is emitted only from objects at high temperatures (above 1500 K). Stars and galaxies are formed from condensation of cold gas and hence in their early stages do not emit visible light—the same is also true as they die having burnt their available fuel. Even at these lower temperatures they still emit radiation, but of a longer wavelength. At temperatures of 30–300 K they radiate in the infrared region (100–10 micrometres) but this radiation is absorbed by the Earth's atmosphere.

The infrared astronomical satellite (IRAS) was designed to survey the sky in this wavelength band. It was a joint USA, Dutch, UK project costing about £100 million. The 1-tonne satellite was launched into a high (900km above the Earth) circular, polar orbit with a revolution time of about 100 minutes. The plane of the orbit was kept perpendicular to the line to the Sun. The satellite included a 60-cm diameter primary telescope mirror that focused radiation onto an array of 62 infrared detectors in the focal plane. These detectors were sensitive to bands of radiation centred on 10, 20, 60 and 100 micrometres. The telescope and the detectors were cooled by a reservoir of liquid helium. The initial charge was 70 kg and the mission was planned on the calculation that this would last 200 days. The telescope was extremely sensitive and could record infrared radiation at the level of 10^{-19} Wcm^{-2}; this corresponds to the radiation received from a 60 W light bulb just after it is switched off at 10 000 miles from the telescope.

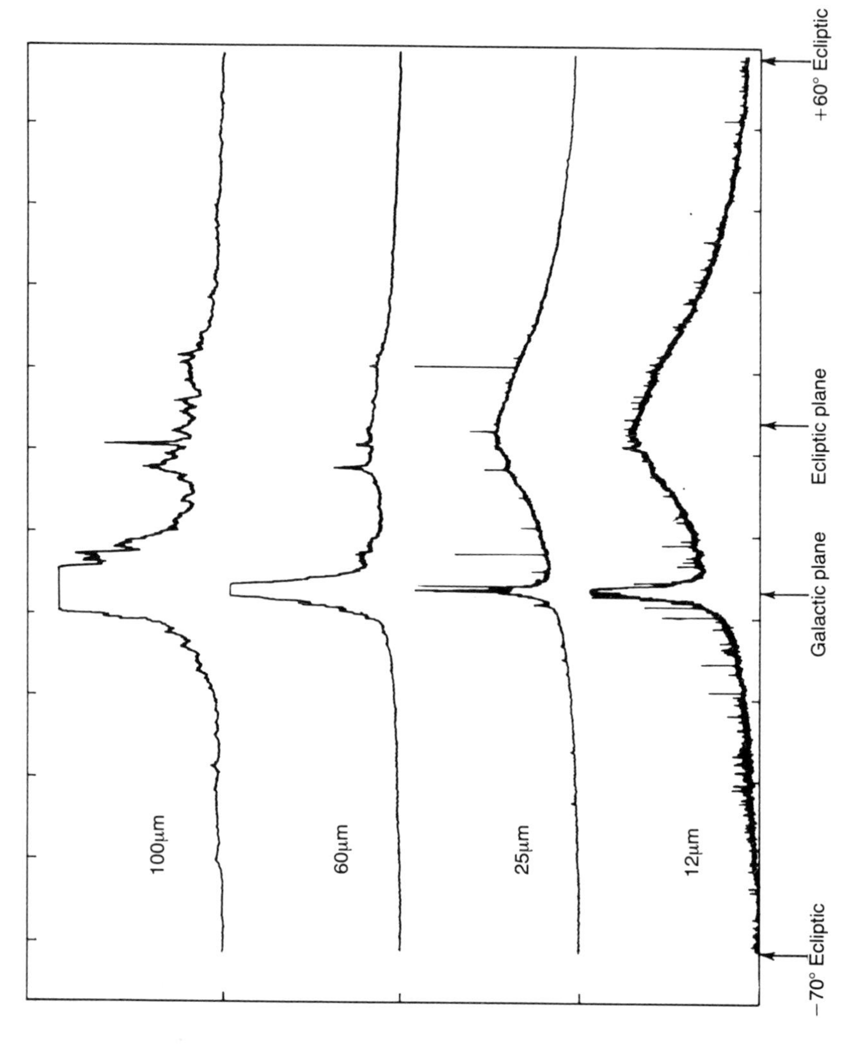

Figure 3 A typical signal as a function of time seen from four of the IRAS detectors. Point sources appear as single 'spikes' which can then be confirmed by subsequent detection in other detectors and on subsequent passes.

The data gathered by the telescope were stored on a tape deck on-board the satellite and was then transmitted at 1 Mega bit/second back to a ground station at RAL twice per day. The helium supply in practice lasted for 301 days and a total of 2×10^{11} bits (equivalent to 4000 times the content of the Bible) was gathered. The survey scanned more than 95 per cent of the sky and included multiple coverage to ensure that the signals observed were from real stellar objects rather than spurious background.

Figure 3 shows a typical trace of the signals received as a function of time. The image of a given stellar object moved across the focal plane as the satellite made one orbit of the Earth. On the next orbit, the image was displaced about half way across the focal plane and hence the object was viewed three times in a period of 200 minutes. The detectors in the focal plane were duplicated so that a given object was seen six times. The data were then processed to average the observed signal and to remove objects that had moved in the sky. This processed data were reconstructed on a computer display to give images of the sky that could be compared with those from other telescopes, such as ground-based optical telescopes.

The experiment was a great success and several important discoveries have already resulted. The complete analysis of the data will take many years but below I give a brief survey of the main results already achieved.

(1) Complete survey in the infrared: A catalogue of 250 000 point sources has been produced—compared with the previous observed number of about 1000.

(2) New galaxies: A number of objects have been seen that emit extremely strongly (about 10^6 times that of the Sun) in the infrared but that had not been seen in the visible region. Some of these have now been detected on ground-based telescopes. They are believed to be galaxies at an early stage of their creation.

(3) New stars: A number of proto-stars—stars at an early stage in their life—have been detected. They are probably only 10^5 years old compared with a lifetime of the universe of more than 10^{10} years. Their temperature is (30–300) K and they can be seen in a cloud of colder gas.

(4) Cirrus clouds: Much of the sky is filled with sparse interstellar dust at low temperature and on the IRAS sky images this has the appearance of 'cirrus clouds'.

(5) New planetary systems: Several examples of stars have been seen with a surrounding ring of infrared radiation which is probably from small particles at low temperature. The first example seen was Vega where the temperature of the particles is 88 K and confirmatory observations have now been made with other telescopes. These results are interpreted as planetary systems in an early stage of their evolution, prior to this there were no known examples of an analogue of our planetary system.

(6) The centre of our Galaxy: The centre of our own Galaxy is not visible with normal optical telescopes, because of the absorption of the dust clouds within our Galaxy. IRAS has now seen this region in the infrared.

(7) Solar system dust bands: IRAS has discovered three well defined dust bands within our planetary system—one is in the plane, one above and one below. These dust bands are believed to be the result of collisions between asteroids.

(8) New comets: IRAS was able to detect comets and to measure their trajectories by their motion in the sky between successive observations. Seven were detected and detailed information was obtained on them.

Overview of complexity and search for simplicity

I have demonstrated the complexity of the RAL programme. We have more than 4000 users and we interface to more than 100 committees. We deal with experimental equipment that goes from the micrometre size of microelectronics to CERN's new LEP accelerator of 27-km circumference. We are concerned with temperatures from 10^{-3} K to 10^{7} K and power densities of 10^{-19}W cm^{-2} (IRAS) to 10^{16} W cm^{-2} (lasers). We have more than 500 computers and 80 000 high density magnetic tapes. Our programme spans 10^{-16} cm to 10^{29} cm.

So if the complexity is easy to demonstrate where is the simplicity? There is behind all of our studies the dream of all physicists that we will uncover an explanation of all known phenomena within the universe in terms of one single force. We even hope that once we have discovered the ultimate truth it will prove to be elegant and simple—perhaps even simple enough to write on the front of a T-shirt. I say the front because if we succeed we will all wish to carry it before us.

Professor Jack Meadows

Jack Meadows is at the University of Leicester, where he is Head of the Departments of Astronomy and the History of Science, the Primary Communications Research Centre, and the Office for Humanities Communication.

He spent his National Service at Cambridge, studying Russian. He then went to Oxford, where he read Physics followed by a doctorate in Astronomy.

After a spell in the United States, at the University of Illinois and the Mount Wilson and Palomar Observatories, and in Scotland at the University of St Andrews, he moved to London. He worked there in the Department of Printed Books and Manuscripts at the British Museum, obtaining in the same period a postgraduate qualification in the History and Philosophy of Science at University College London. He subsequently moved to the University of Leicester, where he has been able to pursue all his interests.

In terms of communication, he is currently a member of various advisory committees, such as the Library and Information Services Council, which advises the relevant Minister. He is also involved, in an editorial capacity, in a number of relevant journals.

How scientists communicate: simplifying complexity

Just as children acquire language unthinkingly along with their mothers' milk, so scientists tend to absorb the mechanics of writing research papers unconsciously along with their supervisors' sour grapes. Consequently, the process of recording and transmitting research information is seen by most scientists as straightforward—apart from arguments with editors and referees. In fact, the construction of an acceptable research paper reflects the agreed view of the scientific community on what constitutes science. A study of the way papers are constructed at any point in time therefore tells us something about the scientific community at that time. Studies over a period of time can tell us whether the community's view of science changes with time. From this viewpoint, the scientific paper can be seen as a kind of archaeological artefact. The serried ranks of bound volumes on library shelves can be dug out and investigated to provide some kind of picture of the environment in which they were created.

A brief, though hardly adequate, definition of science is that it is cumulative knowledge. This definition does have the virtue of emphasizing the central importance of preserving scientific information. Since the early days of the scientific journal in the seventeenth century, the basic medium for preservation has been print on paper. Quite clearly the current trend is to transfer some of the past activities of printed journals to electronic media. The really interesting question is whether the implicit assumptions of the scientific community, that printed journals have learned to fulfil, can be satisfied by the present means of electronic handling. Investigating the structure of scientific papers helps make some of the factors involved explicit, and so ensures that they can be taken into account when planning new developments.

Development of the printed journal

In order to discuss how journals reflect change with time, it is necessary to find titles which cover a long timespan. The obvious choice—equivalent to an archaeological sequence going back to the early Stone Age—is the *Philosophical Transactions of the Royal Society*. It provides a useful basis for discussing changes with time: the main points can be readily confirmed by reference to other titles.

Looking at the titles pages first, the early ones are greatly cluttered with information. As the years pass they simplify and the amount of information is abbreviated to essential elements only. The same condensation, though in a different form, is evident in the papers published. Initially, they are simply printed letters—they have little organization, no recognizable references, etc. Gradually, a certain amount of organization appears along with a few references, given in a very fragmentary form. At the same time, the date at which the paper was read to the Society starts to be added. Structured papers, in the sense we would use this description today, appear in the nineteenth century. Much more informative references (though still inadequate by modern standards) are grouped in footnotes, rather than appearing in the body of the text. This standardization is aided by the systematic imposition of refereeing on submitted papers. The date of receipt of a paper begins to be added to supplement the date of reading. Finally, in the twentieth century and especially after the Second World War, all the paraphernalia of a modern paper appear—the abstract at the beginning, the detailed references at the end, the reprint reference at the head of the paper, and so on. (For a fuller description of these changes, see ref. 1.)

All these changes can be interpreted in terms of developments in the scientific community. For example, one of the key questions in science has always been: who has priority for a particular advance? Initially, when scientific papers were essentially letters, the date of writing would do to establish priority. However, this lays itself open to obvious possibilities for abuse; so, as the scientific community developed, it was laid down instead that the date at which the material was presented orally to the Society should count. This date was then attached to the printed paper. By the nineteenth century, the number of scientists had grown to the point where there could be a long delay before a paper came up for reading. This again raised possibilities of abuse. A paper might be read more quickly if submitted to a smaller society, and so permit priority to be claimed by a scientist who was not actually first in the field. For this reason, the date of receipt came to replace the date of reading. However, as the scientific community continued to grow, and competition to increase, the feeling that it was actually necessary to publish first grew. Journals were therefore put under pressure to publish papers more quickly. (One way of noting this is to compare the interval between the date of receipt of a paper and the cover date of the journal issue: though the latter, of course, says nothing about the actual date of appearance of the issue.) Ultimately, this led to letters' sections with faster publication in journals and to the establishment of letters' journals with rapid publication times.

What these changes reflect is the continuing attempt by the scientific community to control controversy over priority claims despite increasing competition caused by a rapidly expanding community. (That these attempts at regulation have had some success is suggested by Merton's finding[2] that the incidence and bitterness of priority disputes has actually decreased with time.) Similar attempts at interpretation can be applied to almost any aspect of scientific papers. I shall select for brief comment here three which bear particularly on the question of transfer to electronic media.

Information in titles

As the volume and level of specialization of scientific knowledge has increased, so has the need for efficient information retrieval. This has been reflected in the history of journals by the appearance of abstracts and, correspondingly, of abstracts journals. However, the first line of retrieval is nearly always via the title of a paper. It makes sense therefore to examine whether there have been any interesting changes in the titles of research papers with time. Table 1 records some changes in the number of words in titles since the Second World War.[3] We can divide such words into two groups—those which are of little use for retrieving informtion (such as 'and' or 'the') and those which are sufficiently information-laden to be used for retrieval. The latter sort are called 'substantive' words, and their numbers are listed in the final column of the table. What is apparent is that

Table 1 *Changes in the titles of research papers with time*

Journal	*Year*	*No. of words per title*	*No. of substantive words per title*
Analytical Chemistry	1947	8.6	5.6
	1962	9.8	6.7
	1973	11.6	8.0
Journal of Organic Chemistry	1947	7.2	4.7
	1962	7.4	4.9
	1973	10.2	6.6
Lancet	1947	6.2	4.2
	1962	8.2	5.5
	1973	9.0	6.3
Philosophy	1945–50	4.8	2.8
	1960–63	5.0	3.0
	1971–74	4.5	2.8

both the total number of words and the number of substantive words in titles have increased for all three scientific journals, but not for the philosophy journal.

What seems to have happened is that, while the number of scientific papers in circulation has gone up rapidly in the post-war years, so has the amount of information provided by authors in the titles to their papers. hence, the ease of retrieval via titles has remained much the same. Research papers in the humanities have been under less pressure, and so have been less likely to show a systematic increase in the information content of their titles. The substantive words in a title relate to the keywords typically used in retrieval from computerized databases. The ability of titles to keep up, in information terms, with the growth of science implies that the efficiency of computerized search procedures should not be greatly impaired. However, an across-the-board study of journals suggests that not all disciplines in science have kept up equally well. Hence, there might be enhanced retrieval problems in particular areas. Equally, there are some subjects outside the sciences where the information content of titles has also increased. These might prove to be as efficiently organized as the sciences for computerized information retrieval.

Citation analysis

Having started with the titles at the beginning of papers, we now move to the lists of references with which they end. I mention them simply because their study via citation analysis has long been recognized as a method of casting light on the nature of the scientific community. I began work in this area in the 1960s looking at what came to be called the "immediacy effect". Clearly, since the amount of literature is increasing with time, there will be fewer references to older literature, because there will be less older literature to cite. However, in many sciences recent papers are cited even more heavily than would be expected from the rate of growth of scientific literature. This excess amount constitutes the immediacy effect. Derek Price explained the effect in terms of the existence of a "research front".

From these relative simple early steps, citation analysis has developed to the stage where it is actually used as an aid in considering the allocation of funding to researchers or facilities. That is to say, study of the structure of scientific papers has in this case moved from a passive examination of the scientific community to an active involvement in its development. Everyone who is seriously concerned with the use of citation analysis in this way is aware of the dangers of pushing the data too far. It is worth quoting Garfield's comments on this point.[4]

"Citation frequency is a measure of research activity, or of communication about research activity. The measure is a sociometric device. In itself, the number of citations of a man's work is no measure of *significance*. Like one scale on a nomogram, it must be used along with other scales to obtain anything useful or meaningful, particularly if the object of the evaluation is in any way qualitative."

One of the relevant points about citation analysis, in terms of our present theme, is that it would hardly be possible without the use of computers: the amount of data and the number of operations involved are too large. Thus such activities as the mapping of science via co-citation strengths, currently under way at ISI in the United States, depend on the production of references in machine-readable form. From this viewpoint, we can say that the references in scientific papers have already made the transition to the new medium: they are better handled via computers than in their printed form. (For comments on all these developments, see ref. 5.)

Retrieving information from text

One of the characteristic changes in the text of scientific papers over the past three centuries has been a trend towards a formal and impersonal style of writing. In the seventeenth century, a scientist might write: "As my old friend, John Smith, told me . . ." His peer in the twentieth century puts it differently: "According to Smith (1984) . . ." Correspondingly, the readability of scientific papers has come in for increasing condemnation as the years have passed. To quote one editor commenting on his experiences:[6]

"Dr Smith submitted to one of my journals a surprisingly well-written, well-prepared manuscript, his previous manuscripts having been poorly written, badly organized messes. After review of the new manuscript, I wrote "Dr Smith, we are happy to accept your superbly written paper for publication in the *Journal*." However, I just couldn't help adding: "Tell me, who wrote it for you?"

Dr Smith answered: "I am so happy that you found my paper acceptable, but tell me, who read it to you?"

The obvious solution is to provide courses or manuals on good science writing. This has been done, but this is only a partial solution. It is also necessary to ask why scientific papers have become more 'unreadable', for we are dealing with a long-term trend. The answer to this is tied up with how scientists read a scientific paper. It is rare for them to read it straight through from beginning to end. They are more inclined to browse—scanning particular parts of the paper to derive items of interest to them as rapidly as possible. This strategy, which seems to be related to the large number of items available for reading, is one cause of the more highly structured papers that are characteristic of modern scientific journals. But it also appears to be a major factor in the scientific style of writing. What is difficult to read as continuous text may be just right for scanning and rapidly extracting information. There is a, no doubt apocryphal, story of a major learned society publisher of scientific journals which tried to improve the quality of the English in the papers it published. The experiment was discontinued, allegedly, when Japanese readers (who bought as many copies as British readers) complained that they found the improved English more difficult to read. It is certainly true that attempts to alter the style of scientific papers can only be done within the limits permitted by the needs of the scientific community.

The transfer of text to electronic media raises in an acute form the importance of structuring in scientific papers. Browsing in such papers is often a non-linear activity: the reader jumps backwards and forwards in the text, being guided by the physical layout of the paper. What equivalent guidance can be provided for a person browsing at a computer terminal? Will scientists accept different ways of being routed through a paper?

Conclusion

In the early 1960s, Medawar asked—is the scientific paper a fraud? He meant by this that the average scientific paper is a highly sanitized and rationalized account of the actual research process. Hence, he expected the answer—yes, it is a fraud. It should be evident by now that a more defensible answer would actually be—no! Scientific papers are as carefully constructed as any archaeological artefact: they reflect the requirements of the contemporary scientific community, changing as it changes. It follows that, if you wish to transfer the same information to another medium you must recognize these community requirements. Otherwise, your attempts at transfer are likely to fail. Finally, the best way of discovering these requirements is by analysing papers in printed journals, since their long history offers some guarantee that they do fit the needs of the community.

Acknowledgments

The substance of this paper was first given at this conference in 1985, it was subsequently given, in a modified form, to a meeting of the London section of the Institute of Information Scientists.

References

1. Katzen, M.F., 'The changing appearance of research journals in science and technology: an analysis and a case study', in Meadows, A.J. (Editor), *Development of Science Publishing in Europe,* pp. 177–214, Elsevier Science Publishers, Amsterdam, New York & Oxford (1980).
2. Merton, R.K., 'Singletons and multiples in scientific discovery', *Proc. Am. Phil. Soc., 105,* 470–486 (1961).
3. Buxton, A.B., and Meadows, A.J., 'The variation in the information content of titles of research papers with time and discipline', *J. Doc. 33* 46–52 (1977).
4. Garfield, E., 'Citation frequency as a measure of research activity and performance [Current Contents, 1973]' in Garfield, E. (Editor), *Essays of an Information Scientist,* Volume 2, pp. 406–408, ISI Press, Philadelphia (1977).
5. Garfield, E., *Essays of an Information Scientist,* ISI Press, Philadelphia, Volume I (1977); Volume II (1977); Volume III (1980).
6. Day, R.A., *How to Write and Publish a Scientific Paper,* ISI Press, Philadelphia (1983).

Professor John Maynard Smith FRS

John Maynard Smith is Professor of Biology at the University of Sussex.

His qualifications span Engineering and Biology, with a first degree in Engineering (BA, Cambridge) and seven years in aircraft design, followed by BSc Zoology from University College London. His academic career as Lecturer and Reader at University College was continued at the University of Sussex as Professor of Biology from 1965, and he became the first Dean of the School of Biological Sciences.

He was elected Fellow of the Royal Society in 1977, and Foreign Member of the American Academy of Arts and Sciences. In 1982 he was elected Foreign Associate, US National Academy of Sciences.

His publications include numerous scientific papers and eight books on a wide variety of biological topics including evolution and ecology, honing his treatments with mathematical insights including the developing tools of modelling, stability theory, and theory of games.

The unfolding organism and the lesser known work of Alan Turing

The advent of space travel has presented us with a new problem: how can we recognize a Martian as being alive? I suggest that the defining characteristic of a living organism is that it should be 'adaptively complex'. By saying that a thing is complex we mean, among other things, that it would take a lot of information to describe it: a sphere 1 metre in diameter made of lead is not complex. But that is not enough: the surface of the Atlantic Ocean at any instant in time would take an enormous amount of information to describe, yet we would think of it as random rather than complex. Certainly we would not be tempted to think that it was alive. In addition to requiring a lot of information to describe, a living organism is 'adaptive', in the sense that its parts, or organs, serve 'functions'. Legs are for walking, eyes for seeing, kidneys for getting rid of waste substances, and so on. We can go further and state that the parts of organisms serve a particular kind of function: they serve to ensure the survival of reproduction of the organism itself.

If, then, we land on Mars and observe an object which is hard to describe, and which has parts with clear functions, we will think that it is alive. We might be fooled if we observe a steam engine or a computer—that is, an object which is not alive but was made by something that was. But we should be able to tell the difference. The parts of artefacts serve functions, but those functions are not usually concerned with the survival or reproduction of the object itself.

The next question is: how do adaptively complex objects come into existence? Biologists since Darwin have a simple answer to this question. Any population of entities which have the three crucial properties of multiplication, heredity and variation are likely to evolve until they become adaptively complex. Multiplication means simply that one entity can give rise to several. Heredity is defined in Figure 1. It requires that there be multiplying entities of more than one kind. Heredity is then the property that like begets like: that A's give rise to A's, B's to B's, and so on. Variation is then simply the property that heredity is not perfect: occasionally an A gives rise to a C.

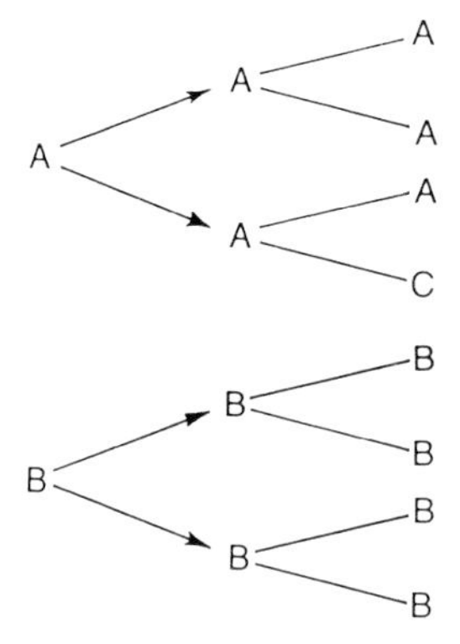

Figure 1 The definition of heredity. In replication, A's give rise to A's, and B's to B's. Variation requires that an occasional error is made; for example, an A gives rise to a C.

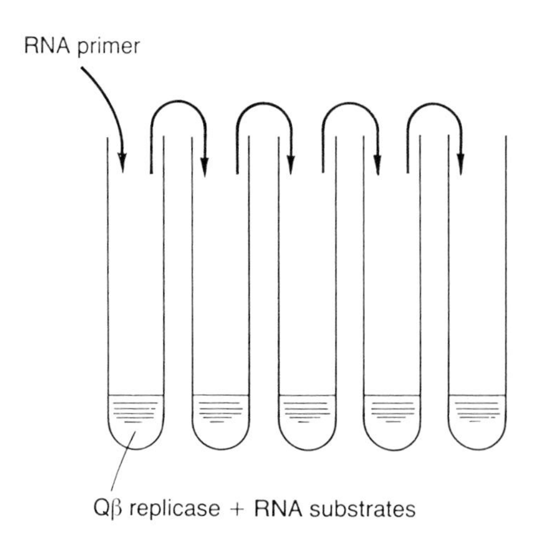

Figure 2 The evolution of RNA molecules in test tubes. After ref. 1.

Are these properties necessary and sufficient for the evolution of adaptive complexity? They are certainly necessary, but they are not sufficient, as can be seen from the following experiment.[1] A test tube contains the four bases (adenine, cytosine, uridine, guanine) from which the polymer, RNA, is made, together with an enzyme (Qβ replicase) that can replicate RNA. If this tube is seeded with an RNA molecule, that molecule will be replicated. After many copies have been made, a drop is taken from the tube and used as a seed for a second tube, and so on indefinitely (Figure 2). Replication is not accurate, so the base sequences of the population of molecules change. However, some sequences are replicated faster than others and thus become commoner. This 'evolution by natural selection' does not continue indefinitely, but only until an 'optimal' sequence has been reached. This optimal sequence is some 230 bases long, and is such that the molecule bends back on itself to form a hairpin structure. Evolution then stops.

What this experiment shows is that natural selection between replicating entities does produce evolutionary change, producing a highly improbable result in a relatively short time. But it does not lead to an indefinite increase in complexity. For that to occur more is needed. There has to be an 'accessible' and 'selectively favoured' path leading to ever-increasing complexity. By accessible, I mean only that each step in the path, from parent to changed offspring, be of a kind that can arise by chance mutation, and by selectively favoured that each step leads to an increase (or, at least, not a decrease) in chances of survival and reproduction.

Presumably the environment in the experiment just described is so simple that no such path exists: no more complex molecule would do any better.

Animals, however, are not merely strings of bases. They are three-dimensional objects. These three-dimensional objects develop anew in every generation. How they do so is the subject of the rest of this essay, but I must emphasize that the answers are not fully known.

I suggest that there are two fundamentally different kinds of processes involved, which I will refer to as 'jigsaws' and 'waves'. By jigsaws I mean structures whose shape is determined by the shapes of the pieces that fit together to make them. In biology, the pieces are molecules. The sequence of bases in DNA determines (by a rather well understood process) the sequence of amino acids in proteins. In turn, this sequence determines the way the protein will fold up, and hence its three-dimensional shape. Still larger scale structures can then be made by fitting together one or several kinds of proteins, as the pieces of a jigsaw are fitted together. Figure 3 shows a virus that is made by this process of 'self-assembly' of genetically programmed proteins. (The process is a little more complicated than this, but not much more).

It seems certain, however, that larger structures are not made like this. The skull of an elephant is not the same shape as a human skull, but this is not because it is made by fitting together pieces of a different shape. Some analogy other than that of a jigsaw is needed. I believe that the natural analogy is that of a wave.

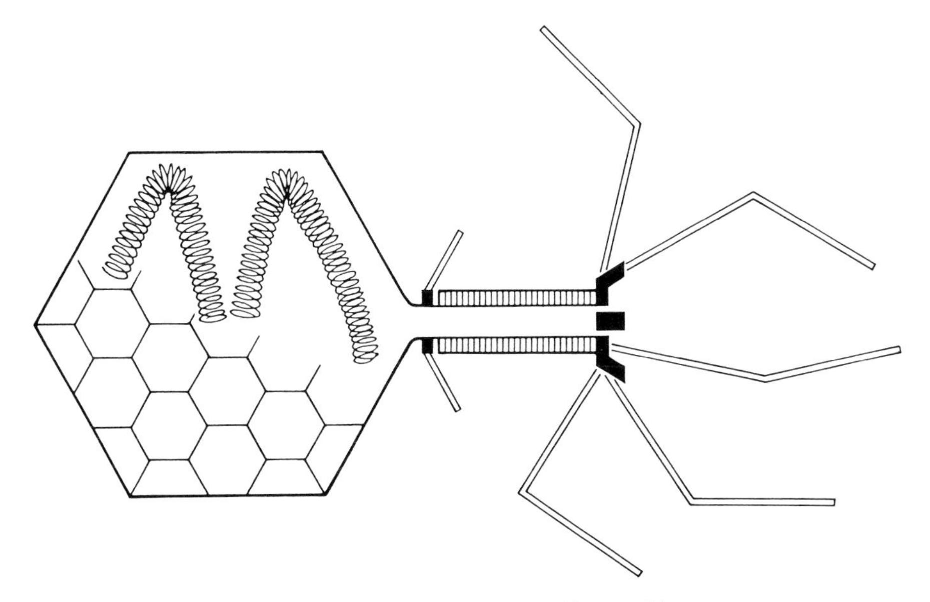

Figure 3 Diagram of a virus that develops by self-assembly.

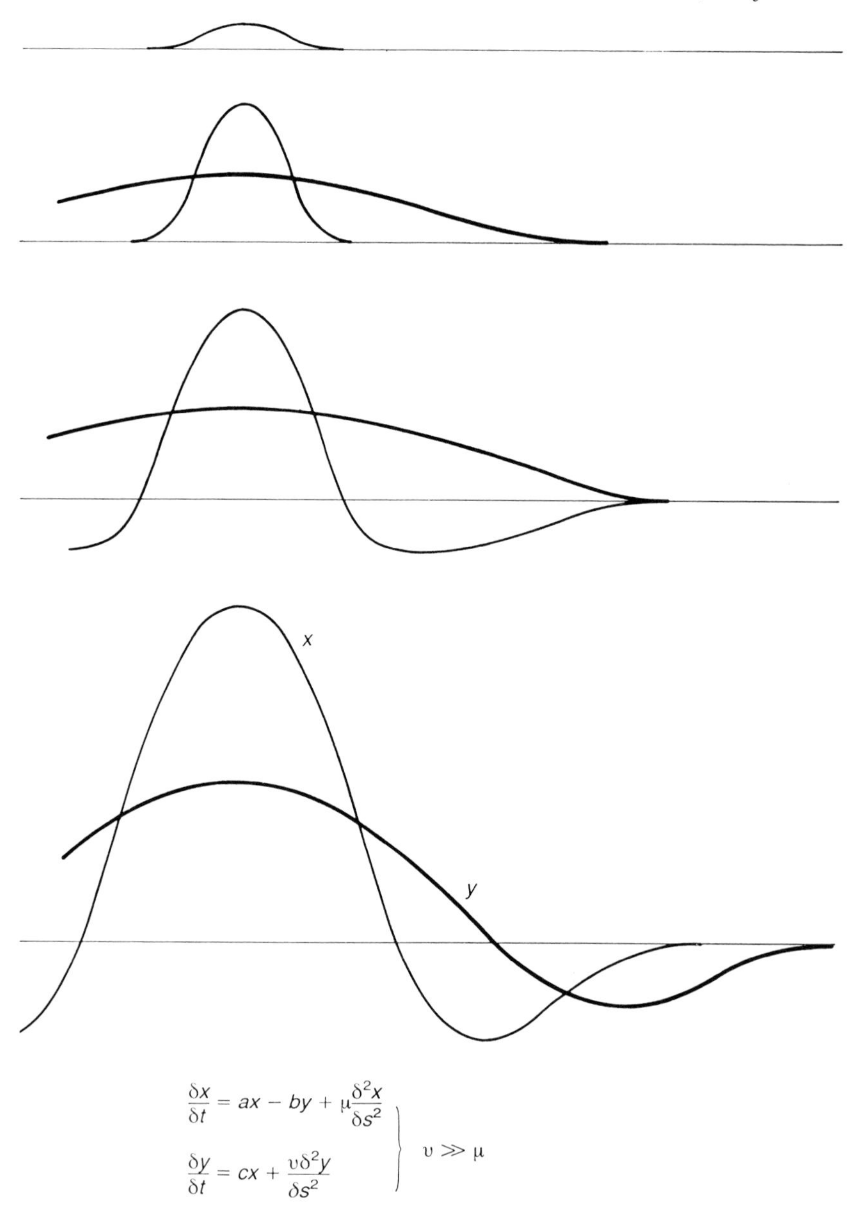

Figure 4 Turing's[2] theory of chemical morphogenesis. There are two morphogens, X and Y, concentrations x and y. X stimulates its own synthesis, and the synthesis of Y. The presence of Y causes the destruction of X. Both substances diffuse, but Y diffuses faster than X. The result is the development of a standing wave.

If we drop a stone in a pond, a series of circular waves will spread out. Their shape does not depend critically on the shape of the stone. Nor does it depend at all on the shape of the water molecules that compose it; the waves would be much the same shape if the pond was made of alcohol. A wave is a simple example of a shape arising that does not depend on the shape of the component parts.

That such processes might be important in biology was first suggested by Turing,[2] in a paper which may play a role in biology comparable to the role played in computing science by his paper on universal computing machines. In essence, he supposed that two chemical substances, A and B, which he called "morphogens", together with the substrates from which they can be synthesized, were free to diffuse in a two-dimensional sheet. He further supposed that autocatalytic reactions occurred: hence the system was far from thermodynamic equilibrium. He showed that, for particular values of the diffusion and reaction rates, a wave of concentrations would develop, as shown in Figure 4. Jabotinsky's reaction, shown at this meeting by Prigogine, in an actual realization of this idea, depends on a rather more complicated set of reactions than were imagined by Turing. Embryologists are still debating the significance of 'reaction-diffusion' mechanisms of the kind proposed by Turing. Most would probably agree that pattern-forming processes of this kind occur, but that they involve more than simple chemical reaction and diffusion.

Let me now describe two ways in which ideas of this kind have been used to describe developmental processes. The first is Stern's[3] idea of "prepattern" and "competence". The crucial experiment is illustrated in Figure 5. Stern argued that normal and "achaete" flies shared a common "prepattern" indicating where bristles ought to form, but that achaete cells lacked the "competence" to respond to this prepattern. An essentially similar explanation was offered by Maynard Smith and Sondhi[4] for variations in bristle patterns produced by artificial selection (Figure 7). Their results differed from Stern's in two ways. They produced flies with bristles at new sites, not present in normal, 'wild-type', flies, indicating a 'submerged peak' of the prepattern (Figure 7*b*), and in other cases they observed flies whose bristle pattern could only be explained if the prepattern itself had changed (Figure 7*c*).

In both these studies, it is easy to imagine the prepattern as consisting of a standing wave of the kind envisaged by Turing. Wolpert[5] has offered a rather different model of development (Figure 6). He supposes that a monotonic gradient is set up, which provides "positional information" to the cells in a sheet. For a two-dimensional pattern, he supposes that two intersecting monotonic gradients exist. The cells then "read off" their positions, and react accordingly. Wolpert's model places less of the onus of pattern formation on field processes (in effect, only the establishment of a

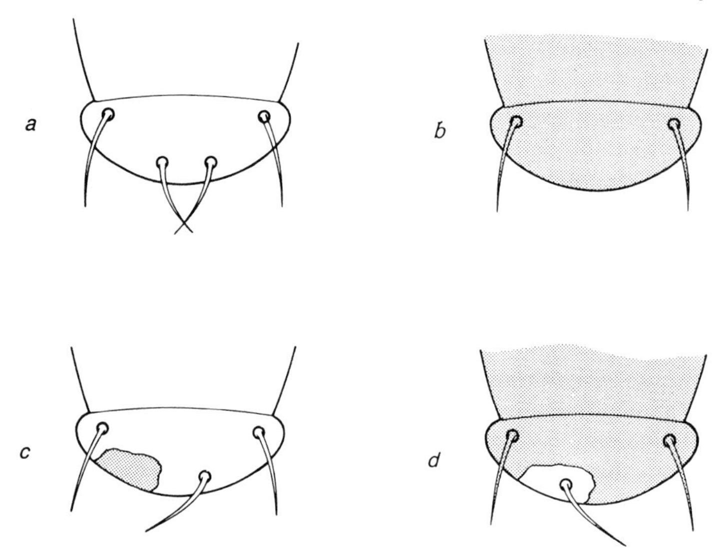

Figure 5 Stern's[3] experiment on the mutant "achaete". Normal flies (*a*) have four bristles on the scutellum. Genetically achaete flies (*b*) have two bristles. Normal flies with a patch of achaete cells where the bristle should be (*c*) lack the bristle. Achaete flies with a patch of normal cells where the bristle should be (*d*) have the bristle.

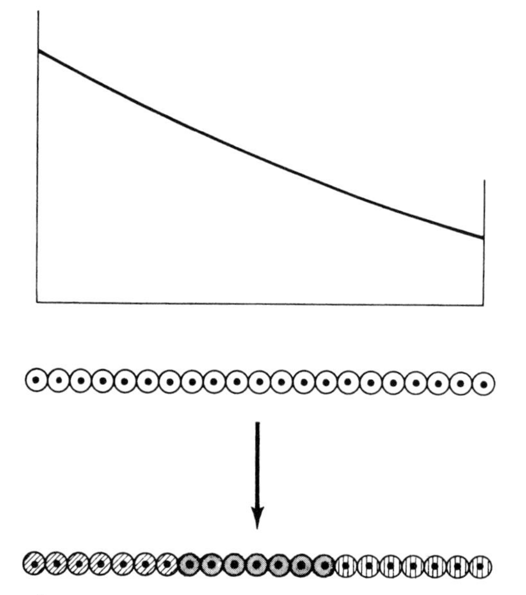

Figure 6 Wolpert's[5] concept of positional information. The information is provided by the value of a monotonic gradient. Cells locally respond in different ways to different values.

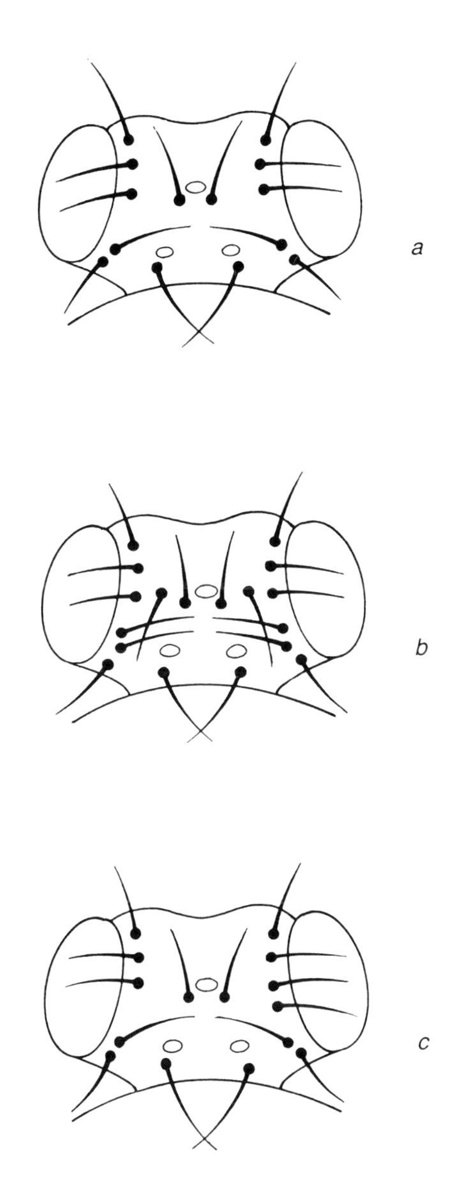

Figure 7 The top of the head of some fruit flies (*Drosophila*) produced by artificial selection, together with an interpretation of the patterns of bristles. *a*, A normal fly; *b*, selection has raised the level of 'competence', so that some bristles are duplicated, and an additional bristle has appeared; *c*, a fly in which the prepattern has changed on the right-hand side. After ref. 4.

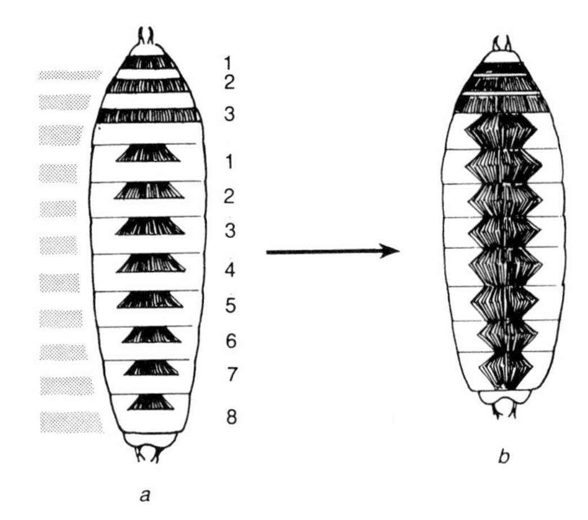

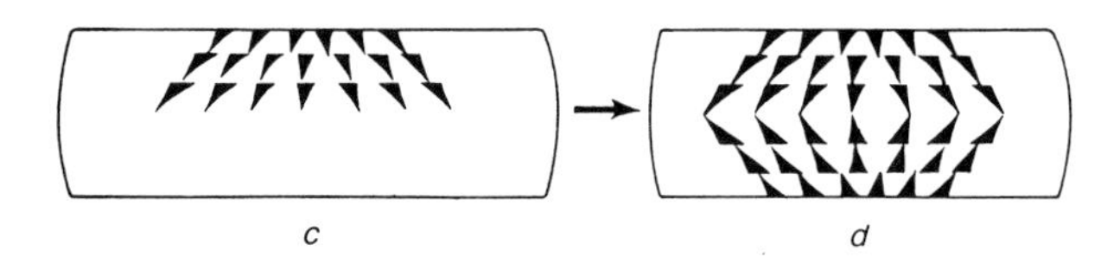

Figure 8 A mutant affecting larval segmentation in *Drosophila*. *a*, A normal larva, with three thoracic and eight abdominal segments; shading on the left indicates regions that are deleted in the mutant. *b*, A mutant larva. *c*, An abdominal segment of a normal larva, showing the peg-like bristles. *d*, An abdominal segment of a mutant larva. After ref. 6.

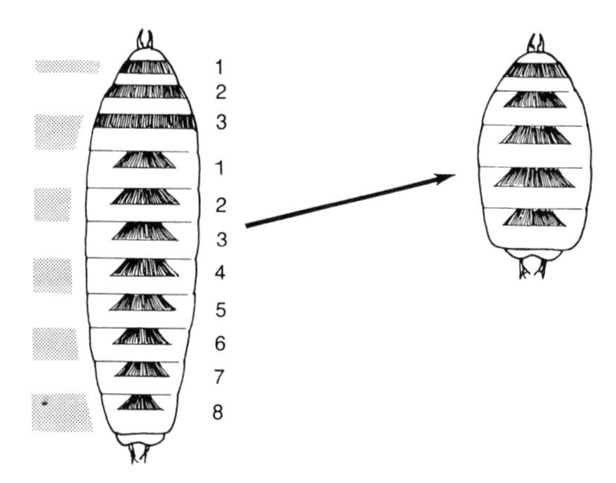

Figure 9 As Figure 8, showing a different mutant.

gradient), and more on the genetic apparatus of the individual cells that respond to the gradient.

More recently, hopes of rapid progress in study of development have come from more detailed genetic analysis. Particularly exciting is the work of Nusslien-Volhard and Wieschaus[6] on segment formation in larval *Drosophila*. They asked how many genes are there such that, if they are mutated, the result is a change in the pattern of larval segmentation. It has turned out that there are remarkably few such genes (probably less than 20). More remarkable still, they alter the segment pattern in relatively few, highly specific ways. Two of these classes of mutants are shown in Figures 8 and 9. In the mutant in Figure 8, the posterior part of each segment is missing, and the anterior part has been duplicated in reverse polarity. In the mutant in Figure 9, every alternate segment has been deleted. The challenge now is to think of a generative process, with relatively few parameters (set by genes), such that it produces the wild-type pattern, and also produces the characteristic mutant patterns when the parameters are changed. Several suggestions have been made which come close to meeting this challenge.

What is new is the possibility of identifying and sequencing the genes concerned with specific developmental processes. I do not think that, by itself, this will tell us how development works, because I do not think that organisms are molecular jigsaws. We must also understand the long-range processes that are going on. Those processes are controlled by genes, but merely knowing the base sequence of a gene will not tell us what they are. As so often in science, we are faced with a particle–wave duality, with a digital–analogue duality. Genes are digital, but patterns are analogue. We have to understand both.

References

1. Orgel, L.E., "Selection *in vitro*", *Proc. Roy. Soc.*, *B205*, 435–442 (1979).
2. Turing, A.M., "The chemical basis of morphogenesis", *Phil. Trans. Roy. Soc.*, *B237*, 37–72 (1952).
3. Stern, C., "Two or three bristles?", *Am. Sci.*, *42*, 213–247 (1954).
4. Maynard Smith, J., and Sondhi, K.C., "The arrangement of bristles in *Drosophila*", *J. Embryol. Exp. Morphol.*, *9*, 661–672 (1961).
5. Wolpert, L., "Positional information and the spatial pattern of cellular differentiation", *J. Theor. Biol.*, *25*, 1–47 (1969).
6. Nusslein-Volhard, C., and Weischaus, E., "Mutations affecting segment number and polarity in *Drosophila*", *Nature*, *287*, 791–801 (1980).

Acknowledgement

We gratefully acknowledge permission granted by: William Heinemann Ltd. to include extracts from two chapters from *Superforce* by Paul Davies; Psychic News for Figure 3 in the chapter by Dr. Alan Gauld, *Ghosts in the Machine*. Professor Jack Meadows' paper How scientists communicate: simplifying complexity also appears in *Journal of Information Science*.